Novel Sustainable Concepts and Technologies in Civil Engineering (NSCTCE-2022)

IOP Conference Series: Earth and Environmental Science
Volume 1149

Vamanjoor, India
22-25 September 2022

Editors:

Chandrasekhar Bhojaraju
Ganesh Babu Kodeboyina
Claudiane M. Ouellet-Plamondon

ISBN: 978-1-7138-7434-8
ISSN: 1755-1307

Printed from e-media with permission by:

Curran Associates, Inc.
57 Morehouse Lane
Red Hook, NY 12571

CURRAN ASSOCIATES INC.
proceedings
.com

Some format issues inherent in the e-media version may also appear in this print version.

Institute of Physics
Dirac House, Temple Back
Bristol BS1 6BE UK

Phone: 44 1 17 929 7481
Fax: 44 1 17 920 0979

techtracking@iop.org

Additional copies of this publication are available from:

Curran Associates, Inc.
57 Morehouse Lane
Red Hook, NY 12571 USA
Phone: 845-758-0400
Fax: 845-758-2633
Email: curran@proceedings.com
Web: www.proceedings.com

TABLE OF CONTENTS

ADVANCES IN CIVIL ENGINEERING

Effect of data normalisation in estimating wave overtopping discharge parameter of semicircular breakwater using ANN and Random Forest. 1
A Shankara Krishna, Vishwanath Mane, Subba Rao, Manu

Ultra-high Performance Concrete as a Sustainable Structural Composite..... 16
Ganesh Babu Kodeboyina, Lakshmi Thotakura, Deepti Avirneni, Chandrasekhar Bhojaraju

Sustainable strategies for monument protection - A panoramic view of the possibilities and pitfalls..... 25
Ganesh Babu Kodeboyina

Numerical simulation of a floating-type point absorber wave energy converter..... 33
Sharath Srinivasamurthy

Numerical Investigation the Effects of Cone Diameters on the Flow Pattern and Separation Efficiency in a Cyclone Separator 39
Seyed Soheil Mousavi Ajarostaghi, Seyed Sina Mousavi, Chandrasekhar Bhojaraju

Numerical Analysis of Double Stack Blade Savonius Wind Turbine with Secondary Blades 55
Seyed Soheil Mousavi Ajarostaghi, Seyed Sina Mousavi, Chandrasekhar Bhojaraju

BUILDING MATERIALS AND STRUCTURAL ENGINEERING

Probabilistic Analysis of RC Buildings Based on Incremental Dynamic Analysis 71
Merin Mathews, B R Jayalekshmi, Katta Venkataramana

Mechanical and Microstructure Properties of Plain and Blended Cement Structural Lightweight Aggregate Concrete Using Sintered Fly Ash Aggregates..... 77
Pankaj Dhemla, Prakash Somani, Raj Kumar Satankar, B L Swami

Rubberized Geopolymer Mortar and Concrete: A Comprehensive Review. 89
P K Arjun Raj, D Sarath, Praveen Nagarajan, A. P. Shashikala

Study on the effectiveness of prefabricated cage system reinforcement in columns..... 103
T Thejarathnam, M H Prashanth, Imran Kuttagola

Numerical investigation of the hybrid reinforced concrete beam using GFRP bars..... 115
Aditya Kumar, M H Prashanth, Imran Kuttagola

ENVIRONMENTAL, TRANSPORTATION AND GEOTECHNICAL ENGINEERING

A Comparative Study on Dynamic Response of Buildings Resting on Coir and Rubber mat Reinforced Soil Bed 122
M V Sreya, B R Jayalekshmi, Katta Venkataramana

Fillers Influence on Hot-Mix Asphalt Mixture Design and Performance Assessment 127
Vinay Hosahally Nanjegowda, M N Rathankumar, N Anirudh

Author Index

Novel Sustainable Concepts and Technologies in Civil Engineering (NSCTCE-2022) IOP Publishing
IOP Conf. Series: Earth and Environmental Science 1149 (2023) 011001 doi:10.1088/1755-1315/1149/1/011001

PREFACE

Civil engineering has played a pivotal role in shaping the world we live in today. With the constantly evolving world, the need for sustainable development has become more crucial than ever before. The Civil Engineering International Conference on "Novel Sustainable Concepts and Technologies in Engineering (NSCTCE 2022)" aims to address the current trends and put new research and advances into practice. As engineers, it is essential to prioritise public safety, and welfare and uphold the ideals of sustainable development without jeopardising the future generations' ability to do the same.

The NSCTCE-2022 abstract proceedings offer a suitable space for trending themes with well-balanced material and a manageable quantity of submissions. The program covers a wide range of civil engineering disciplines, including structural engineering, geotechnical engineering, water resources engineering, transportation engineering, and environmental engineering.

The event was graced by 12 keynote speakers from India and overseas, who presented their exceptional research in a range of civil engineering topics, making it a valuable resource for both academics and practicing engineers. The conference received an overwhelming response, with about 86 abstracts being submitted, out of which 40+ were chosen for full-length presentations. The selected papers will be published in the journal 1OP: Earth & Environmental Sciences, further contributing to the dissemination of knowledge.

We would like to express our sincere appreciation to all the authors who showed keen interest in the three-day NSCTCE-2022 conference. We are also incredibly grateful to the management of SJEC, Mangalore, sponsors, and exhibitors for their support, which the conference's success was only possible if it had been successful. We hope that NSCTCE-2022 has encouraged fruitful discussions, sparked new ideas, and provided a platform for sharing research and experiences, ultimately contributing to the betterment of society through sustainable development.

Editors

Chandrasekhar Bhojaraju,
Ganesh Babu Kodeboyina,
Claudiane M. Ouellet-Plamondon

Our conference committee included

CHIEF PATRON
Most Rev. Dr Peter Paul Saldanha Bishop, Diocese of Mangalore & President - SJEC

PATRONS
Rev. Fr Wilfred Prakash D'Souza, Director-SJEC
Rev. Fr Alwyn Richard D'Souza, Asst Director-SJEC

CONFERENCE CHAIR
Dr Rio D'Souza, Principal, SJEC

ADVISORY COMMITTEE
Dr. Venkata Ramana Akkaraju, Chairman-NMPT
Prof .Prasad Bhaskaran, IIT Khrg
Prof. Mehdi Dehestani, BNUT-Iran
Prof. K.V.L.Subramanyam, IIT H
Prof. Pradeep Kumar Ramanacharla, IIIT H, Director CBRI, Roorkee
Prof. K Ganesh Babu, Former Director CBRI, Roorkee
Prof. C Lakshmana Rao, IITM
Er. Nagesh Puttaswamy, Ultratech
Prof. Subbarao, NITK
Prof. Ananth Ramaswamy, IISC
Dr. Suresh D S, Chairman, ISTE, Karnataka Section

TECHNICAL COMMITTEE
Chair
Prof. Claudiane Ouellet-Plamondon ETS Montreal, Canada
Members
Prof. Katta Venkataramana, NITK
Prof. Bibhuti Bhusan das, NITK
Prof. Manu, NITK
Dr. Tejo Vikash Bheemasetti, The University of Arizona
Er, Devdatta Bose, TCE
Prof. Rajeevan B., GCE
Prof. Arun Kumar Bhat, NITTE
Dr. Hemaraju P, GITAM
Dr. Radheshayam B, AECOM India Pvt Ltd.
Dr. Anand V R, VCET Puttur
Organizing Chair
Dr. Yajnheswaran B, Prof. & HOD - Civil, SJEC
Convener
Dr. ESTR Chandrasekhar B, Associate Prof., SJEC
Organizing Secretary
Mr. Manjunath B, Assistant Professor, SJEC

IOP Conf. Series: Earth and Environmental Science 1149 (2023) 011001 doi:10.1088/1755-1315/1149/1/011001

Treasurers
Mr. Prashant K, Mr. Chitharanjan M & Ms. Pranamya K Assistant Professors, Civil Dept, SJEC
Publication Co-ordinators
Mr. Govind Krishna, Ms. Vishika K, Ms. Sreya MV
Ms. Reena Rodrigues and Mr. Bhojaraja M Assistant Professors, Civil Dept, SJEC
Publicity Co-ordinators.
Mr. Ajeeth B, Mr. Sachin U and Mr. Rohit P Velgas
Assistant Professors, Civil Dept, SJEC

Photos of the Event

Sponsors in association with ACCE Mangalore, IGBE, ISTE, IEI.

- Type of peer review: Double anonymous
- Conference submission management system: Morressier
- Number of submissions received: 25
- Number of submissions sent for review: 25
- Number of submissions accepted: 14
- Acceptance Rate (Submissions Accepted / Submissions Received × 100): 56 %
- Average number of reviews per paper: 2.00
- Total number of reviewers involved: 18
- Contact person for queries:

Name: Chandrasekhar Bhojaraju
Affiliation: Associate Professor
Email: chandrasekhar.b@sjec.ac.in

Effect of data normalisation in estimating wave overtopping discharge parameter of semicircular breakwater using ANN and Random Forest.

A Shankara Krishna[1], Vishwanath Mane[1], Subba Rao[2] and Manu[3]

1 Ph.D. Research Scholar, Dept. of Water Resources and Ocean Engineering, National Institute of Technology Karnataka, Surathkal.
2 Professor, Dept. of Water Resources and Ocean Engineering, National Institute of Technology Karnataka, Surathkal.
3 Associate Professor Dept. of Water Resources and Ocean Engineering, National Institute of Technology Karnataka, Surathkal.

E-mail: shankarakrishnaa.217wo008@nitk.edu.in; vishwanathmane@gmail.com; subbarao@nitk.edu.in; manunitk@gmail..com;

Abstract. Breakwaters are the structures constructed in the coastal areas to maintain calm inside the port or prevent beach erosion. Semi-circular Breakwater (SCB) is an innovative type of Breakwater made of hollow caisson on a base slab with or without perforations. In this study, the wave overtopping discharge parameter of an SCB is estimated using Artificial Neural Network and Random Forest. The data is collected and used in the current research from an experimental investigation conducted in the Wave Mechanics Laboratory of the Department of Water Resources and Ocean Engineering (WROE), NITK Surathkal. Using this experimental data, the ANN and Random Forest models are developed for the prediction of the wave overtopping discharge parameter of an SCB. The performance of the models is evaluated using different statistical parameters. Data with and without normalisation are used separately to check the effect of normalisation in the prediction of wave overtopping discharge parameter using ANN and Random Forest. From the results, it is found that ANN gives better results when the data is normalised. The performance of Random Forest is independent of the data normalisation.

1. Introduction

The use of sea routes for transportation is more because the cost of transport is cheaper than other modes in the case of long-distance transportation. Therefore, the development of ports and harbours is essential. Maintaining a calm condition inside the port is vital for the smooth functioning of work. For this purpose, breakwaters are constructed so that the wave activity is minimal inside the ports. Breakwaters are also built to prevent erosion along coastal areas. Different types of breakwaters are used in the field, such as Rubble mound breakwater, vertical wall type breakwater, composite breakwaters, innovative type breakwaters, etc. [1]. According to the site conditions, the type of Breakwater is selected. Semi-circular Breakwater is one of the innovative types of breakwaters used in the coastal region. The nomenclature and abbreviations used in the present paper are listed in the table given below.

Table 1. Nomenclature and abbreviations used.

Nomenclature		Abbreviations	
Incident wave height	H_i	Semicircular Breakwater	SCB
Water depth	D	Root Mean Square Error	RMSE
Time period	T	Nash-Sutcliffe efficiency	NSE
Radius of structure	R	Correlation Coefficient	CC
Percentage perforation	p	Scatter Index	SI

1.1. Semi-Circular Breakwater (SCB)

SCB is a hollow caisson structure. It consists of a hollow semi-circular caisson made of prestressed concrete and a base slab placed on a rubble base foundation. The construction material required for SCB is less since it is hollow. The shape of the structure gives good stability and a better aesthetic view. Figure 1 shows the schematic diagram of SCB.

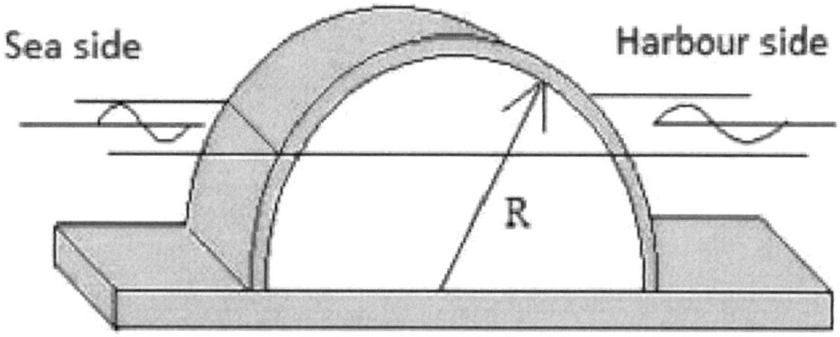

Figure 1. Schematic diagram of SCB.

SCB can be classified in two ways

1. Submerged and Emerged type

In the submerged type, the structure will be under the water; in the case of the emerged type, the crest will be above water level.

2. Permeable and Impermeable type

It can be classified into four: Solid type with no perforation, the permeable type having perforation at the front, back, or both sides together. The permeable type having perforations at the front is used the most in the field.

Many research works were conducted on the semicircular breakwater[2]. Hegde et al. (2012) carried out a study on perforated SCB models[3]. Perforations are given on the face where waves are striking. They have considered different water depths (d), wave heights (Hi), wave periods (T), depth parameter (d/gT^2), incident wave steepness (Hi/gT^2) and relative spacing S/D (S = c/c spacing of perforations, D = diameter of perforations) to evaluate the wave run-up, wave run-down, and reflection coefficient on the seaside. It was noted that when incident wave steepness or percentage perforation or depth parameter increased, the reflection coefficient decreased. When incident wave steepness or depth parameter increased, relative run-down (Rd/Hi) was shown to decrease. The relative run-up decreased as the depth parameter was increased. Hegde et. Al. carried out study on stability of SCB [4]. Emerged SCB with perforation on the seaside is considered for the study. Sliding stability of structure is mainly depends on the structure weight. Hence in this study critical weight required for the structure in various wave and structural conditions are found. A dimensionless stability parameter ($W/\gamma H_i^2$; W-Critical weight or minimum weight required for stability, γ-Specific weight of water) is used to analyse the stability. From the study it is found that critical weight increases with increase in the water depth at the site.Sasajima et al. (1994) introduced the concepts of the different types of SCB. The incident wave energy dissipates while the wave runs over the curved surface of the SCB and causes partial wave reflection.

1.2. Artificial Intelligence

Artificial Intelligence is widely used in various fields in the present day to tackle various real-life problems. Machine learning is a type of artificial intelligence where machines learn from past behaviour or data and predict future outcomes. Machine learning is widely employed, from social media advertisements on our mobile devices to contemporary combat weaponry. It has many applications in coastal and ocean engineering [5][6][7].

ANN is a technique inspired by the biological nervous systems and it is used in various fields of engineering [8][9][10]. They can be trained and used for the prediction of different parameters. Van Gent et al. (2005) constructed a neural network to forecast the wave overtopping at various coastal structures [11]. It was observed that neural network prediction was fairly accurate in the range of high overtopping discharge. Zanuttigh et al. (2016) used ANN to estimate wave overtopping discharge at coastal and harbour structures for different wave conditions considering various geometries [12]. They have added additional datasets to focus on extreme and tolerable overtopping discharges (discharge $q<10-6$ and $q>10-3$). They optimised the existing ANN (Zanuttigh et al. (2014)) to predict small discharge values. Kuntoji et al. (2020) used Particle Swarm Optimisation based ANN and Support Vector Machine to predict wave transmission on tandem breakwaters[13]. Using different statistical parameters performance of each model is calculated. The hybrid PSO-SVM model accomplished better results contrasted with PSOANN. Harish et al. (2015) developed a Support Vector Machine (SVM) and a hybrid model of Particle Swarm Optimisation (PSO) with SVM to estimate the severity of non-reshaped berm breakwater damage. PSO is used for the optimisation of kernel parameters. PSO-SVM model with polynomial kernel function came out as the best model[14].

The ensemble is a machine learning technique that combines the number of base models to improve the result. There are three different ensemble methods: bagging, boosting and stacking[15][16]. Random Forest is an example of a bagging technique[17]. This technique is excellent in regression and classification problems. Sun et al. (2019) developed a new beetle random forest model (BRF) by combining beetle antennae search (BAS) and random forest (RF) algorithms[18]. They have used it to establish the relationship between the uniaxial compressive strength of Rubberised concrete with its important parameters. The results concluded that BAS effectively tuned RF. Hence hyperparameters were procured with the help of BAS. The developed BRF model predicted the uniaxial compressive strength of Rubberised concrete with a high correlation coefficient of 0.96. Campos et al. (2021) successfully used random forest (RF) to decide the best wave forecast between deterministic and ensemble products of the National Center for Environmental Protection (NCEP)[19]. The National Data Buoy Center (NDBC) buoy data was used to train the supervised learning classifier, and the RF model performance was evaluated as a function of forecast time. Graw et al. (2020) used a Random Forest regressor to predict the value of seafloor density of other locations from the data of an area, which is well known[20]. The study concluded that density predictions match expected lateral density variation for ocean depth and low uncertainties for the majority of the seafloor.

2. Methodology

The methodology followed in the present study is shown in figure 2.

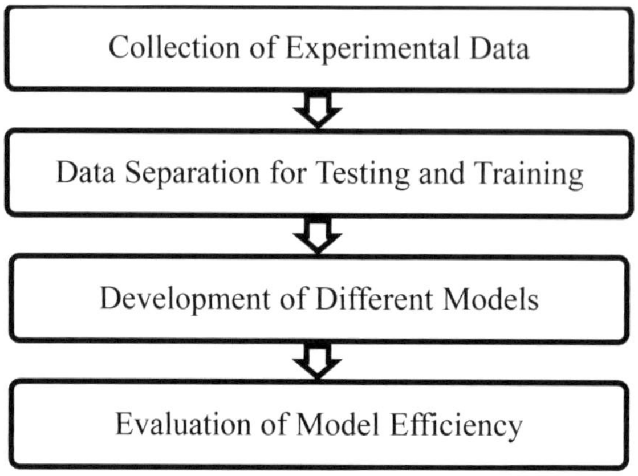

Figure 2. Methodology flow chart of the present study.

An experimental study was conducted on the performance of SCB in a two-dimensional wave flume in the Department of Water Resources and Ocean Engineering, NITK Surathkal. This data is used in the present study and the details are shown in table 2.

Table 2. Experimental variables.

Parameters	Experimental Range	
Wave Specific Parameters		
H_i (m)	0.08, 0.10, 0.12, 0.14, 0.16	
d (m)	0.45,0.425,0.4 (For SCB of radius 0.45 m)	0.4,0.375,0.35 (For SCB of radius 0.40 m)
T (s)	1.4, 1.6, 1.8, 2.0, 2.2	
Structure specific parameter		
R (m)	0.40, 0.45	
p (%)	0, 1.25, 5, 10, 15, 20	
Diameter of perforation(m)	0.016	

The wave flume is 50 metres in length, 0.71 metres in width, and is 1.1 metres deep. The bottom-hinged flap (wave paddle) generates waves in a 6.3 metres long, 1.5 metres wide, and 1.4 metres deep chamber at one end. An induction motor controls the flap. The line diagram of the wave flume is shown in figure 3.

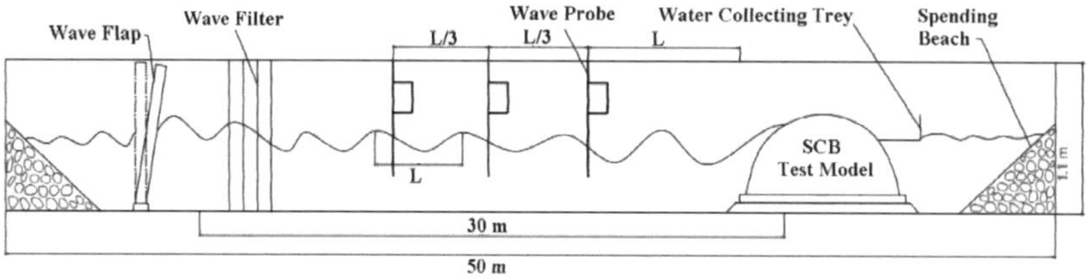

Figure 3. Longitudinal section of wave flume (Not to scale).

The collected data is divided into two sets for testing and training randomly. The percentage of data used for training and testing are 70 and 30, respectively. ANN and Random Forest models are developed after separating the data to predict the overtopping discharge of SCB. MATLAB is used to develop the ANN model, while an open tool called "Orange (version 3.29)" is used to create Random Forest model. After getting the results, the model's efficiency is assessed using variety of statistical parameters such as Root mean square error (RMSE), Nash-Sutcliffe efficiency (NSE), Correlation coefficient (CC) and Scatter index (SI).

3. Development of Model

In this study, 770 sets of data of SCB are considered. It consists of different radii (0.45 and 0.40 m) and percentages of perforations (p=0%, 1.25% ,5% ,10% ,15% ,20%) at different wave heights, wave time periods, and water depths. It is understood that the overtopping discharge is mainly dependent on incident wave steepness (Hi/gT2), depth parameter (d/gT2), crest height parameter (Rc/Hi) and percentage of perforation (p). Hence these four variables are selected as input parameters and the non-dimensional mean overtopping discharge parameter(q/gHiT) is the output. The data is divided into 70-30 ratio, i.e., 539 samples for training and 231 samples for testing[21]. These inputs, output parameters and dividing the data are the same for every model considered in this study. The best model is picked by considering the values of various statistical parameters mentioned earlier (RMSE, NSE, CC, SI).

In this study, two cases are considered.

Case 1: Data without normalisation. In this case, the given data is used as it is obtained.

Case 2: Normalised data is used. In this case, the given data is normalised between 0 to 1 using the following equation [22].

$$X_{norm} = \frac{X - X_{min}}{X_{max} - X_{min}} \quad (1)$$

X_{norm} – Normalised value of X between 0 to 1
X – Value of data point which is to be normalised.
X_{max} – Highest of given data points. X_{min} – Lowest of given data points

3.1. *Artificial Neural Network (ANN)*

ANN is a technique inspired by biological neurons. It is used in different fields to solve different types of complex problems. They can be trained for forecasting future events, Image processing and character recognition, etc. In recent developments, it has been used for speech recognition also. The architecture of ANN consists of 3 layers[23] (i)The input layer-where the data for the network is provided. (ii)Output layer-where the results from the network are obtained. The hidden layer is in between these two layers. For most of the problems, this 3 layer configuration is sufficient. More than one hidden layer is used in highly complex problems. In this study, ANN is used to find the trend between input and output values so that it can be used to predict the output values for problems of similar conditions [24]. Figure 4 shows the network diagram of ANN developed in MATLAB. The network is named 4-6-1, where 4 is the number of input parameters. Six is the number of neurons present in the hidden layer. One is the number of output parameters from the developed ANN model.

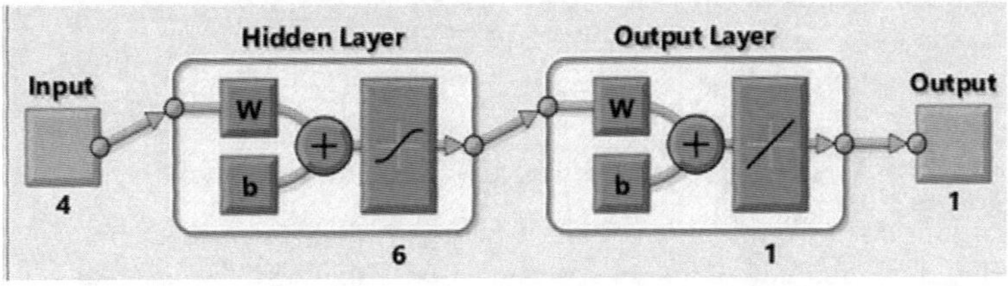

Figure 4. Network diagram of ANN (MATLAB 2019b).

ANN model is developed using MATLAB (2019b). A three-layer ANN feed-forward backpropagation neural network is used in this study and transfer functions are used between the layers of ANN. The transfer function converts the input data into matrices so that it can be readable by the machines. Similarly, it converts the results to make it easy for humans to understand the output. The present study uses 'tansig' and 'purelin' transfer functions between input-hidden and hidden-output layers. The predicted value is compared with the given target value in the feed-forward network training, and the errors are backpropagated. Based on this the weights are adjusted to minimise the error. This process is carried out till the error reaches the given minimum value. In this study, the Levenberg-Marquardt algorithm is used for predicting the overtopping discharge of a SCB. The number of neurons in the hidden layer is varied and the performance is calculated.

3.2. *Random Forest*

Random forest is an ensemble machine learning technique used for classification and regression problems. A decision tree is the basic building block of Random Forest and it constructs 'n' number of decision trees while training. Figure 5 shows the structure of a simple decision tree. For the classification type of problem, the output given by the most number of trees is selected and in the case of the regression problem, the average value of results obtained from each tree is considered. Figure 6 shows the general structure of Random Forest.

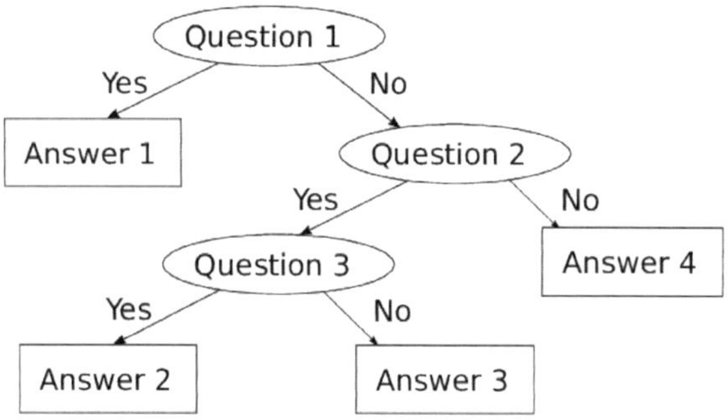

Figure 5. Structure of a simple decision tree [25].

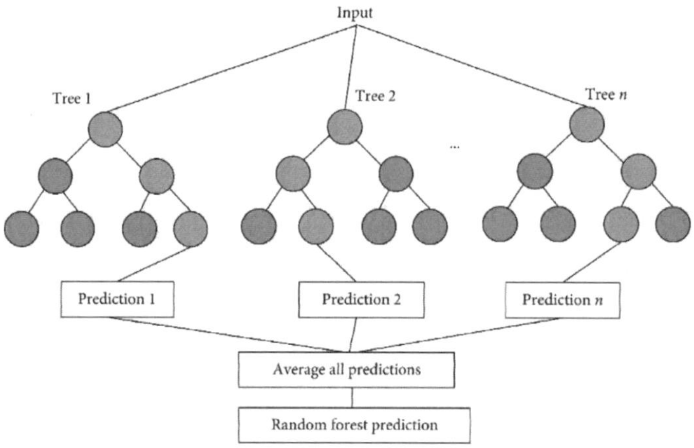

Figure 6. General structure of Random Forest [18].

A new data set is formed from a given data set randomly. The new data set will be a subset of the given data set. Based on this new data set, decision trees are formed and results are obtained. The average value of results obtained from all decision trees are expressed as the final result. The performance is calculated based on this result. An open tool called Orange (version 3.29) is used to develop random forest model. In the present study, the number of trees used to develop the random forest model is 100. The subset will not split below 5 to control the tree's growth.

The performance of each model is calculated by using four statistical parameters. The statistical parameters used in the present study are RMSE, NSE, CC, SI. The equation used to find these parameters are given below [13], [26].

$$\text{RMSE} = \sqrt[2]{\frac{1}{N}\sum_{i=1}^{N}(Pi - Oi)^2} \qquad (2)$$

Where Oi = observed /actual values; Pi = computed values obtained from various models; N= Number of data points. Normally the model with low RMSE is considered as most accurate.

$$\text{NSE} = 1 - \frac{\sum_{i=1}^{N}(Pi-Oi)^2}{\frac{1}{N}\sum_{i=1}^{N}(Oi-Oavg)^2} \qquad (3)$$

O_{avg} = Mean of actual values

$$\text{CC} = \frac{\sum_{i=1}^{N}(Pi-Pavg)(Oi-Oavg)}{\sqrt[2]{\sum_{i=1}^{N}(Pi-Pavg)^2} * \sqrt[2]{\sum_{i=1}^{N}(Oi-Oavg)^2}} \qquad (4)$$

P_{avg} = mean of computed values

$$\text{SI} = \frac{\sqrt[2]{\frac{1}{N}\sum_{i=1}^{N}(Pi-O\ )^2}}{Oavg} \qquad (5)$$

4. Results and Discussion

Results obtained from the different models are discussed in this section. ANN and Random Forest are used for developing the model. ANN is developed using MATLAB 2019b and Random Forest using a tool called Orange (version 3.29). The data is loaded into the model in two ways, with and without normalisation to check the effect of data normalisation in the prediction of wave overtopping discharge parameter. Hyperparameter is a parameter whose value is used to control the learning process. In the present study, hyperparameter tuning is done manually by varying those parameters using the provision given by the tool.

4.1. Analysis of Result Without Data Normalisation

The following result is obtained for ANN in this case. MATLAB 2019b is used to develop the ANN model. The number of neurons in the hidden layer is varied from 1 to 8. The results obtained are shown in Table 3. The best result is obtained for 4-6-1 ANN model.

Table 3. Statistical parameter of ANN models for SCB.

Network	RMSE		NSE		CC		SI	
	Train	Test	Train	Test	Train	Test	Train	Test
4-1-1	0.0006	0.0006	0.3710	0.4326	0.6095	0.6582	0.7043	0.7075
4-2-1	0.0005	0.0005	0.5057	0.5132	0.7116	0.7185	0.6244	0.6553
4-3-1	0.0006	0.0006	0.3910	0.4315	0.6259	0.6614	0.6931	0.7082

4-4-1	0.0006	0.0006	0.3774	0.4034	0.6156	0.6367	0.7008	0.7255
4-5-1	0.0005	0.0005	0.5029	0.5039	0.7132	0.7106	0.6261	0.6616
4-6-1	**0.0004**	**0.0004**	**0.6097**	**0.6095**	**0.7810**	**0.7808**	**0.5548**	**0.5869**
4-7-1	0.0006	0.0006	0.3277	0.3996	0.5725	0.6335	0.7282	0.7277
4-8-1	0.0004	0.0005	0.5675	0.5507	0.7542	0.7454	0.5840	0.6269

Table 4 shows the results from all models used in the present study (Data used is without normalisation).

Table 4. Results from all models (Data without normalisation).

Model	RMSE		NSE		CC		SI	
	Training	Testing	Training	Testing	Training	Testing	Training	Testing
ANN 4-6-1 model	0.0004	0.0004	0.6097	0.6095	0.781	0.7808	0.5548	0.5869
Random Forest	0.0002	0.0002	0.9569	0.896	0.9806	0.9497	0.1844	0.3028

From Table 4, it is observed that the result obtained from ANN is not as good compared to Random Forest models. When we vary the number of hidden neurons, it does not show any trend in the results. The performance changes randomly from one model to another because the value to be predicted is very small, which is in the range of 10-6 to 10-3. From Table 4, we can understand that result obtained from the random forest model is good compared to ANN. Since random forest use decision trees as the building blocks, the range of output parameters does not matter for the prediction. The obtained NSE and CC values are close to 1 for training and testing. The Scatter index and RMSE obtained are closer to zero, which indicates the model performance is good.

Figure 7 and 8 show the scatter plots for ANN model 4-6-1 for training and testing, for the prediction of non-dimensional wave overtopping discharge parameter (q/gH_iT) of SCB, respectively. Figure 9 and 10 show the scatter plots of training and testing for the prediction of non-dimensional wave overtopping discharge parameter (q/gH_iT) of SCB from Random Forest.

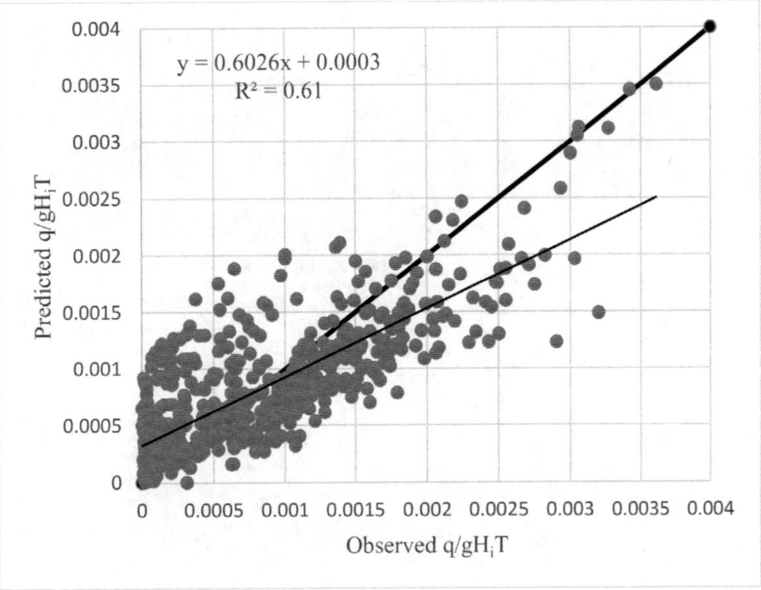

Figure 7. Scatter plot of training of 4-6-1 ANN model.

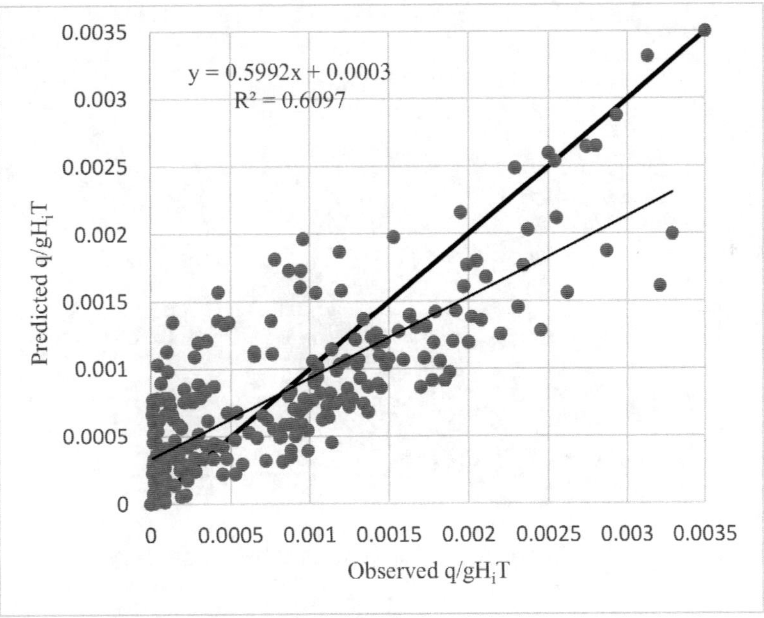

Figure 8. Scatter plot of testing of 4-6-1 ANN model.

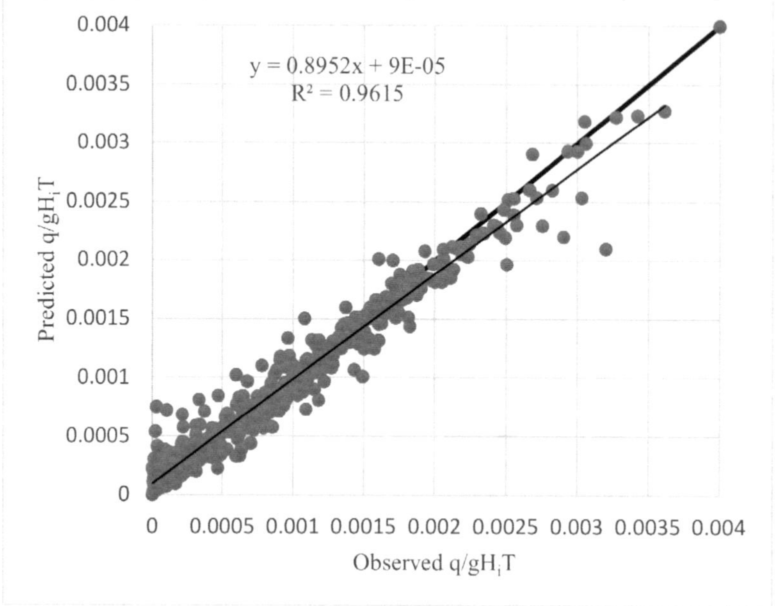

Figure 9. Scatter plot of training of random forest model.

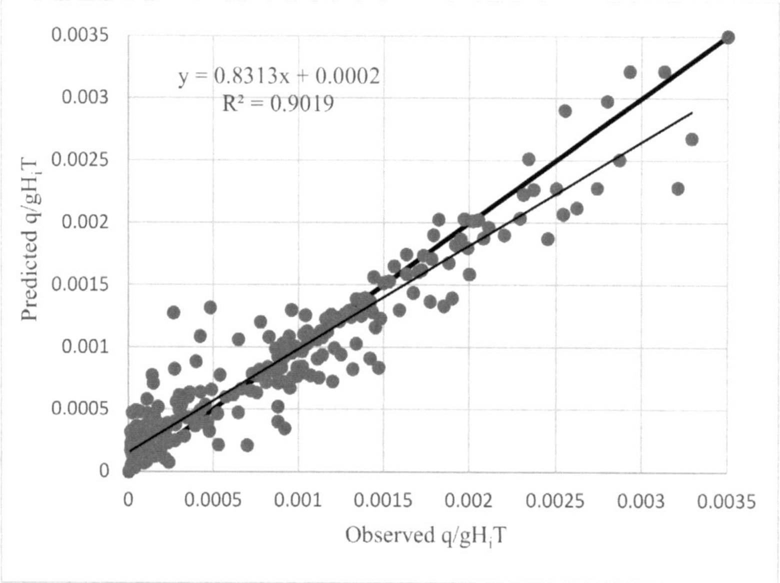

Figure 10. Scatter plot of testing of random forest model.

4.2. Analysis of Result with Data Normalisation

In this case, the data is normalised between 0 to 1. After normalisation it is divided into two sets. Similar to the previous case, 70% of data is used for training, and 30% is used to test the models. Since the data is normalised, the output is normalised value of the non-dimensional wave overtopping discharge parameter (q/gHiT) of SCB, denoted by q* in this paper. The data used for testing and training is the same for all models. The following result is obtained for ANN in this case. MATLAB

2019b is used to develop the ANN model. The number of neurons in the hidden layer is varied from 1 to 11. The results obtained are shown in Table 5. The best result is obtained for 4-6-1 ANN model.

Table 5. Statistical parameters of ANN models for SCB.

Network	RMSE		NSE		CC		SI	
	Training	Testing	Training	Testing	Training	Testing	Training	Testing
4-1-1	0.1617	0.1622	0.3914	0.4437	0.6257	0.6665	0.6929	0.7006
4-2-1	0.1576	0.1612	0.4221	0.4503	0.6511	0.6724	0.6752	0.6964
4-3-1	0.1585	0.1606	0.4155	0.4545	0.6476	0.6780	0.6790	0.6937
4-4-1	0.1452	0.1510	0.5094	0.5181	0.7139	0.7216	0.6221	0.6521
4-5-1	0.1324	0.1408	0.5921	0.5808	0.7696	0.7622	0.5672	0.6081
4-6-1	0.1296	0.1366	0.6092	0.6054	0.7806	0.7782	0.5552	0.5901
4-7-1	0.1421	0.1459	0.5298	0.5499	0.7311	0.7440	0.6090	0.6302
4-8-1	0.1309	0.1396	0.6010	0.5881	0.7756	0.7672	0.5610	0.6029
4-9-1	**0.1240**	**0.1363**	**0.6425**	**0.6071**	**0.8018**	**0.7808**	**0.5311**	**0.5888**
4-10-1	0.1337	0.1427	0.5838	0.5694	0.7649	0.7576	0.5730	0.6164
4-11-1	0.1422	0.1544	0.5297	0.4956	0.7286	0.7044	0.6091	0.6671

Compared to the results obtained from the previous case (where data was used without normalisation), the model developed using normalised data gives better results. In the case of ANN, we can see a proper trend in the result shown in table 5. When the number of neurons increases, the performance of ANN increases up to a particular number of hidden neurons and attain the best performance. Later the performance decreases. A neural network with nine hidden neurons gives the best result with a correlation of 0.8018 for training and 0.7808 for testing. It achieved a RMSE of 0.1240 for training and 0.1363 for testing. ANN gives almost the same performance for training and testing. When compared with other models, the performance achieved by ANN is less. But the results are better when the data was normalised.

The results obtained from different models, in this case, are discussed below. Table 6 shows the results from all models used in the present study (for normalised data).

Table 6. Results from all models (for normalised data).

Model	RMSE		NSE		CC		SI	
	Training	Testing	Training	Testing	Training	Testing	Training	Testing
ANN 4-9-1 model	0.124	0.1363	0.6425	0.6071	0.8018	0.7808	0.5311	0.5888
Random Forest	0.0417	0.0685	0.9595	0.9008	0.9816	0.9523	0.1789	0.2958

From Table 6, it is observed that random forest gives good results with RMSE of 0.0417 for training and 0.0685 for testing. It achieved a correlation of 0.9816 for training and 0.9523 for testing. Compared to the previous case (when data without normalisation is used), the performance is slightly better. But the difference is negligible. Hence, we can say that the normalisation of data has no significant effect on Random Forest. Since the random forest is based on the decision tree, the range of variables does not affect the model's performance. But at the same time, it cannot extrapolate the result. So, if we test the model with an input beyond the training data, the model's performance will reduce. It performs well in the range of trained data (0 to 1 in this case).

Figure 11 and 12 shows the scatter plots for ANN model 4-9-1 for training and testing of the prediction of normalised value of non-dimensional wave overtopping discharge parameter (q/gHiT) of SCB respectively. Figure 13 and 14 shows the scatter plots of training and testing for the prediction of normalised value of non-dimensional wave overtopping discharge parameter (q/gHiT) of SCB from Random Forest.

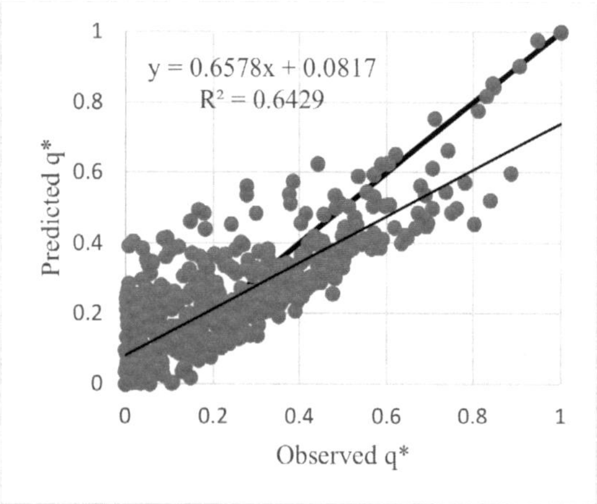

Figure 11. Scatter plot of training of 4-9-1 ANN model.

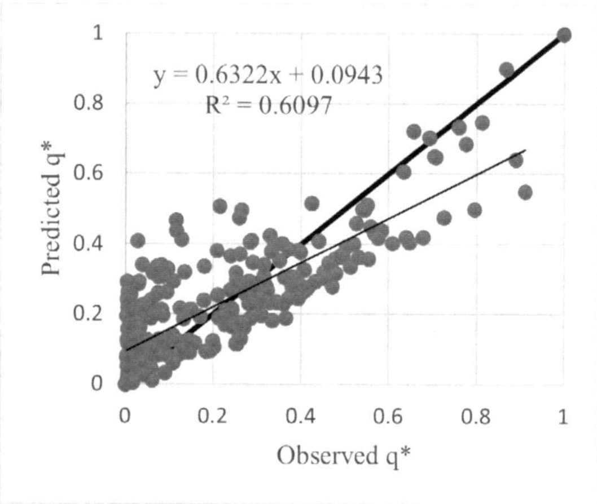

Figure 12. Scatter plot of testing of 4-9-1 ANN model.

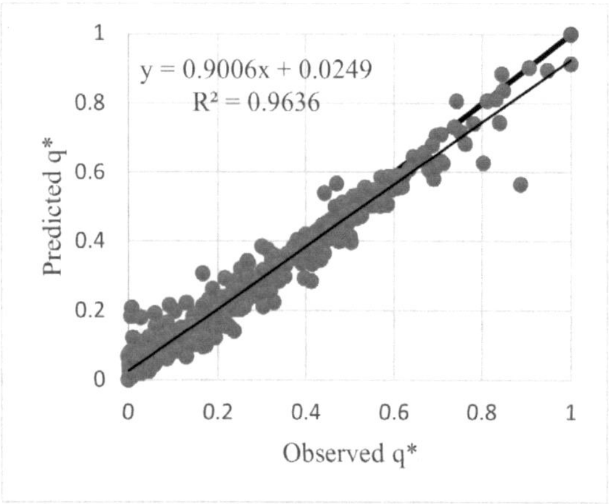

Figure 13. Scatter plot of training of Random Forest model.

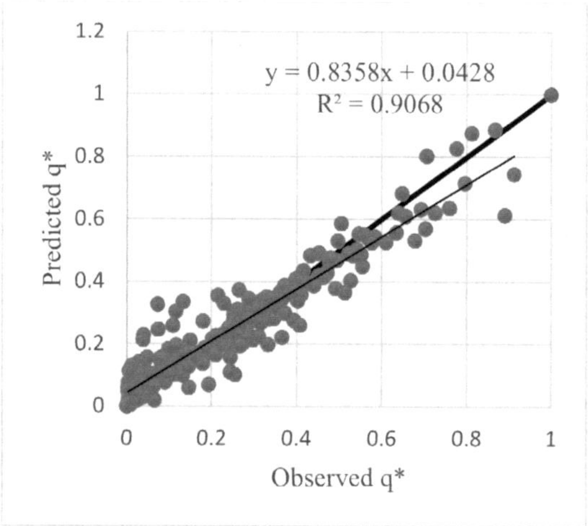

Figure 14. Scatter plot of testing of random forest model.

4.3. Comparison of Performance of All the Models

Table 3 and 5 provide the results from both of the models, which may be compared to see how well they perform. From the table, we can understand that ANN gives better results when the data is normalised. Random Forest gives almost the same result for both original and normalised data. Hence, normalisation of data has no significant effect on Random Forest which is based on decision tree. However, by comparing all the values, we can say that for predicting mean wave overtopping discharge of SCB, using normalised data is slightly better than the data without normalisation. Additionally, it can be inferred that decision tree-based Random Forest model forecast the mean wave overtopping discharge of SCB quite precisely and are hence advised for the same.

5. Summary and Conclusions

5.1. Summary

The soft computing models are developed based on the data acquired from the experimental study on a semicircular breakwater conducted at the wave mechanics laboratory at NITK Surathkal. In this study, 770 sets of data of SCB are considered. It consists data of different radii (0.45 and 0.40 m) and perforations (percentage of perforation = 0,5,10,15,20) at different time period wave heights and depths. It is observed that the overtopping discharge is mainly dependent on incident wave steepness (Hi/gT2), depth parameter (d/gT2), crest height parameter (Rc/Hi) and percentage of perforation. Hence these four variables are selected as input parameters and the non-dimensional mean overtopping discharge parameter(q/gHiT) is the output. The best model is selected based on different statistical parameters like Root Mean Square Error, Correlation Coefficient, Nash-Sutcliffe Efficiency, and Scatter Index.

5.2. Conclusions

Based on the present study, the following conclusions are drawn:

- ANN gives the better result when normalised data is used. Normalised data has no significant effect on the results of Random Forest.
- Among the various ANN models studied for normalised data, the 4-9-1 ANN model gives the best result for the prediction of the mean overtopping discharge parameter of SCB. It achieved an RMSE of 0.124 for training and 0.1363 for testing of the model. And it also obtained CC of 0.8018 and 0.7808 for training and testing, respectively.
- Random Forest achieved good results in the prediction of data. It achieved a CC of 0.9816 for training and 0.9523 for testing. The RMSE, NSE and SI values obtained for this model are 0.0417, 0.9595, 0.1789 for training and 0.0685, 0.9008, 0.0.2958 for testing respectively.

The results show that Random Forest gives the best results for the prediction of wave overtopping discharge parameter. It can be used as an alternative technique for prediction.

Acknowledgements

The authors are thankful to the Director of National Institute of Technology Karnataka, Surathkal, and Head of the Department of Water Resources and Ocean Engineering (Formerly Applied Mechanics and Hydraulics) for the facilities provided and permission granted to publish the result.

References

[1] Rajendra K, Balaji R, and Mukul P, “Review of Indian research on innovative breakwaters, ”Indian J Geomarine Sci, vol. 46, no. 03, pp. 431–452, 2017.

[2] A. V. Hegde and S. M. Naseeb, “Transmission performance of submerged semicircular breakwaters for different radii and submergence ratios,” International Journal of Ocean and Climate Systems, vol. 5, no. 3, pp. 151–161, 2014.

[3] A. V. Hegde, S. Rao, and K. Kumar, “Run-up, run-down and reflection characteristics of semicircular breakwater for varying seaside perforations,” ISH Journal of Hydraulic Engineering, vol. 18, no. 3, pp. 145–151, 2012, doi: 10.1080/09715010.2012.662474.

[4] A. V. Hegde, S. Mohan, J. L. S. Pinho, and Sharhabeel PS, “Physical model studies on the

stability of emerged seaside perforated semicircular breakwaters," Indian Journal of Geo Marine Science, vol. 47, no. 03, pp. 681–685, 2018.

[5] M. C. Deo, "Artificial neural networks in coastal and ocean engineering," 2010.

[6] D. Aleksendric and P. Carlone, "Soft computing techniques," in Soft Computing in Design and Manufacturing of Composite Material, CRC Press, 2015, pp. 39–60. doi: 10.1533/9781782421801.39.

[7] S. Kundapura and A. V. Hegde, "Current approaches of artificial intelligence in breakwaters - A review," Ocean Systems Engineering, vol. 7, no. 2, pp. 75–87, Jun. 2017, doi: 10.12989/ose.2017.7.2.075.

[8] N. S. Hanspal, B. A. Allison, L. Deka, and D. B. Das, "Artificial neural network (ANN) modeling of dynamic effects on two-phase flow in homogenous porous media," Journal of Hydroinformatics, vol. 15, no. 2, pp. 540–554, 2013, doi: 10.2166/hydro.2012.119.

[9] N. Vivekanandan, "Comparison of Artificial Neural Network and Regression Approaches for Evaluation of Hydrodynamic Performance of Quarter Circular Break Water," *International Journal of Advancements in Technology*, vol. 09, no. 01, 2018, doi: 10.4172/0976-4860.1000199.

[10] P. Jain and M. C. Deo, "Artificial Intelligence Tools to Forecast Ocean Waves in Real Time," *The Open Ocean Engineering Journal*, vol. 1, no. 1, pp. 13–20, 2008.

[11] J. W. van der Meer, M. R. A. van Gent, B. Pozueta, H. Verhaeghe, G.-J. Steendam, and J. R. Medina, "Application of a neural network to predict wave overtopping at coastal structures," in *International Conference on Coastlines, structures and breakwaters*, 2005, pp. 259–268.

[12] B. Zanuttigh, S. M. Formentin, and J. W. van der Meer, "Prediction of extreme and tolerable wave overtopping discharges through an advanced neural network," *Ocean Engineering*, vol. 127, pp. 7–22, Nov. 2016, doi: 10.1016/j.oceaneng.2016.09.032.

[13] G. Kuntoji, M. Rao, and S. Rao, "Prediction of wave transmission over submerged reef of tandem breakwater using PSO-SVM and PSO-ANN techniques," *ISH Journal of Hydraulic Engineering*, vol. 26, no. 3, pp. 283–290, Jul. 2020, doi: 10.1080/09715010.2018.1482796.

[14] N. Harish, S. Mandal, S. Rao, and S. G. Patil, "Particle Swarm Optimization based support vector machine for damage level prediction of non-reshaped berm breakwater," *Applied Soft Computing Journal*, vol. 27, pp. 313–321, 2015, doi: 10.1016/j.asoc.2014.10.041.

[15] K. Machova, M. Puszta, F. Barcak, and P. Bednar, "A comparison of the bagging and the boosting methods using the decision trees classifiers," *Computer Science and Information Systems*, vol. 3, no. 2, pp. 57–72, 2006, doi: 10.2298/csis0602057m.

[16] P. Bühlmann, "Bagging, Boosting and Ensemble Methods," in *Handbook of Computational Statistics*, Springer Berlin Heidelberg, 2012, pp. 985–1022. doi: 10.1007/978-3-642-21551-3_33.

[17] L. Breiman, "Random Forests," *Mach Learn*, vol. 45, no. 1, pp. 5–32, 2001.

[18] Y. Sun, G. Li, J. Zhang, and D. Qian, "Prediction of the Strength of Rubberized Concrete by an Evolved Random Forest Model," *Advances in Civil Engineering*, vol. 2019, 2019, doi: 10.1155/2019/5198583.

[19] R. M. Campos, M. O. Costa, F. Almeida, and C. G. Soares, "Operational wave forecast selection in the atlantic ocean using random forests," *J Mar Sci Eng*, vol. 9, no. 3, Mar. 2021, doi: 10.3390/jmse9030298.

[20] J. H. Graw, W. T. Wood, and B. J. Phrampus, "Predicting Global Marine Sediment Density Using the Random Forest Regressor Machine Learning Algorithm," *J Geophys Res Solid Earth*, vol. 126, no. 1, Jan. 2021, doi: 10.1029/2020JB020135.

[21] B. Zanuttigh, S. M. Formentin, and J. W. van der Meer, "ADVANCES IN MODELLING WAVE-STRUCTURE INTERACTION THROUGH ARTIFICIAL NEURAL NETWORKS," *Coastal Engineering Proceedings*, vol. 1, no. 32, p. 69, 2014.

[22] G. S. Kuntoji, S. Rao, Manu, and S. Mandal, "Performance evaluation of ANFIS and SVM model in prediction of wave transmission over submerged reef of tandem breakwater,"

International Journal of Ecology and Development, vol. 32, no. 2, pp. 141–155, 2017, [Online]. Available: www.ceserp.com/cp-jour

[23] F. S. Panchal and M. Panchal, "International Journal of Computer Science and Mobile Computing Review on Methods of Selecting Number of Hidden Nodes in Artificial Neural Network," 2014. [Online]. Available: www.ijcsmc.com

[24] N. Ramesh, A. Hegde, and S. Rao, "Prediction of reflection coefficient of a perforated Quarter Circle Breakwater using artificial neural network (ann)," in *Journal of Physics: Conference Series*, Aug. 2019, vol. 1276, no. 1. doi: 10.1088/1742-6596/1276/1/012006.

[25] U. S. Vallejo, "Occupancy Estimation and People Flow Prediction in Smart Environments," 2017.

[26] G. S. Kuntoji, S. Rao, and S. Mandal, "Performance evaluation of ANFIS and SVM model in Prediction of Wave Transmission over Submerged Reef of Tandem Breakwater," *International Journal of Ecology & Development*, vol. 32, no. 2, pp. 141–155, 2017, [Online]. Available: www.ceserp.com/cp-jour

Ultra-high Performance Concrete as a Sustainable Structural Composite

Ganesh Babu Kodeboyina [1*], Lakshmi Thotakura [1], Deepti Avirneni [1] and Chandrasekhar Bhojaraju[2]

[1] Department of Civil Engineering, Mahindra University, Hyderabad, India.
[2] Department of Civil Engineering, St Joseph Engineering College, Mangaluru, India.

*Corresponding author: ganesh.kodeboyina@mahindrauniversity.edu.in

Abstract. The phenomenal strides in the modifications and the use of supplementary cementitious materials in conjunction with superplasticizers and discrete fiber distributions in concrete made it possible to arrive at the structural material termed as "ultra-high performance concretes" (UHPC) or its extension "ultra-high performance fiber reinforced concretes" (UHPFRC). Its development heralded a new era, opening new vistas in structural configurations and forms that could be a serious contender to structural steel constructions. Primarily, the chemistry and physics of concrete constituents and mixture design philosophy have not changed much, though the intricacies that are often neglected were effectively corrected from time to time. The use of materials ranging from the conventional to nanoparticles that could exhibit superior pozzolanic activity along with microfibers resulted in strengths well beyond 200 MPa. This opens up several avenues in structural configurations that are of high strength and durability. However, this isn't a stipulation of replacing the present-day concrete in all structures with UHPC, but the key is to modulate the structural configurations to suit the needs of the particular application for ensuring sustainability in consonance with the capabilities of UHPC. The paper attempts to look at the possible avenues for such options in materials and structural forms to effectively assure sustainable construction alternates.
Keywords: Sustainability, High Strength, Ultra-high Performance, Environment, Admixtures, Cementitious Materials.

1. Introduction

The advancements in cement and concrete compositions have been so significant that concrete retained its position as the primary construction material over the past couple of hundred years. Even so, the latest trends in the modifications of cementitious concrete compositions exceeded every expectation and made it possible to be a contender in replacing even steel sections. This, termed as the ultra-high performance concretes, is achieved through the synergistic application of plasticizers, supplementary cementitious materials and appropriate grading fillers and aggregates in connection with high-strength cements. Micro-fibrous additions further paved the way for the improving the crack bridging capabilities and the enhancement of ductility of the primarily brittle high-strength cementitious composite. However, all these modifications and additional materials of specific characteristics and compositions, apart from the rigorous processing requirements, naturally increased the cost and availability significantly. Looking at the unit cost alone no one can even remotely imagine that the material UHPC to be a sustainable alternative for the construction application. But its substantial high performance even in highly aggressive environments can offset this, considering the life cycle cost of the structure. It is with this perception that this material UHPC is broadly being evaluated and projected as a sustainable structural composite.

2. Sustainability Perceptions

Sustainability primarily is to achieve a balance between the actions furthering the social and economic aspirations of the society and the resulting environmental impact due to such actions. As is often articulated it rests on the above three pillars of society, economy and environment, with the need to ensure a measured approach which will not compromise the quality of life of the future generations in any way. A very typical example of the dichotomy between development and its impact on the future generations can easily be seen in the adoption or otherwise of nuclear power. The safeguards and consequences necessitating the abnormal maintenance needs of the systems and materials associated due to its adoption is often a passionate topic for discussion for many, well informed or otherwise. The irreversible fact of the depletion of the natural resources that robs future generations the fruits of their availability and probably a more appropriate utilization by them with their superior technological proves is also part of the sustainability paradigm. The above discussions clearly show that sustainability as a concept is essentially subjective proposition for many and trying to quantify even to a limited extent has been difficult. Notwithstanding this a few organizations in different countries have proposed and executed a level assessment in the general building envelope, however subjective they may be.

3. Concrete Composites and UHPC

One of the largest sources of CO_2 emission is always traced to the cement and concrete industry. Naturally, with the ever-increasing need for built space and infrastructure in urban habitation being an undeniable fact, the cement consumption needs to be addressed appropriately for limiting its effects. Reducing the use of cement in the construction activity, either through the use of supplementary cementitious materials or alternative renewable materials or even through the reuse of construction and demolition waste are often considered the best options. However, a better performance of materials in comparison to the normal strength concrete composites used ensures a longer as well as greater service life, making it the subject for discussion in several research efforts. Initial efforts to improve the performance of concrete projected that higher strength concretes at lower water cement ratios (below 0.45) will essentially result in a discontinuous pore structure which will help in reducing the environmental ingress and will lead to a higher performance. The advent of melamine / naphthalene formaldehyde superplasticizers facilitated an easier way to reduce the water cement ratios well below this limit, making it possible to produce concretes of 70 - 80 MPa with ease. The fact that pozzolanic materials like volcanic ash that helped in producing Roman type lime concrete compositions were introduced at this point of time. The use of industrial waste byproducts like fly ash and ground granulated blast furnace slag (GGBS) were also introduced to reduce the consumption of cement and also to achieve a much lower reactivity of the concrete with the environment thus increasing its durability. With the recognition that cements containing GGBS exhibit improved chloride binding capacity, slag-based cements were specifically recommended for use in marine environment. Also, the effective utilization of calcium hydroxide produced during the hydration process, which is deposited mostly in the interfacial transition zone, could be mitigated using these pozzolanic materials. The next era in the development of concrete composites was ushered in with the realization that the addition of silica fume or micro silica with its near nano size particles could produce much higher strength concretes with significant impermeability. The possibility of adopting high efficiency pozzolans like silica fume, metakaolin and other calcined clays that came into existence around this time, in conjunction with superplasticizers, facilitated the production of super high strength high performance concretes in the range of 100 – 120 MPa. These concretes were indeed the mainstay in building most of the skyscrapers of today. Ultimately with of an understanding of these aspects along with an appropriate packing of the filler materials (aggregates) that was hitherto not attempted systematically led to the development of ultra-high strength materials like RPC and MDC. Also, to ensure the bridging of the internal microcracks and impart ductility to such highly brittle matrices, well dispersed microfiber additions came into existence and are termed as UHPCs or UHPFRCs. Figure 1. Presents a broad outline of the various classes of these UHPCs, as perceived by different organizations in the different continents.

Ultra-high performance concrete (UHPC) is a new generation cementitious composites with improved microstructure, high workability and impermeability due to low porosity when compared to the conventional concretes. UHPC has a vast application potential and can be utilized both for engineering and architectural purposes. However, the higher cement requirements of UHPC need a special focus in terms of environmental sustainability. Ultra-high strength concretes need higher amounts of cements which intern leads to high production cost and high emission of CO2 which will indirectly affect the environment, even though some amount of cement is to be replaced by supplementary cementitious materials (SCM) However, such high amounts of pozzolans and fines often require larger amounts of water and to minimize the water content, chemical admixtures have to be incorporated which will also add to the cost. The major advantage is the highest particle packing resulting in a very high strength, that facilitates most economical sections. Some of the possibilities in terms of structural bridge girders and box sections, naturally with due care to stiffen appropriately the slender sections, will ensure a significant lowering of their data weight, a major factor in the design of long span bridges.

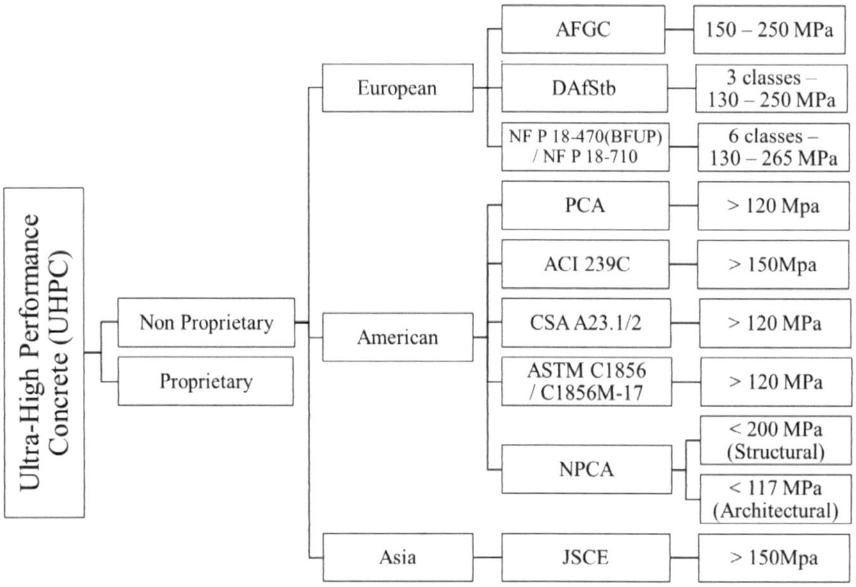

Figure 1. Suggested classification of UHPCs

4. Sustainability Evaluations and Accreditations

Over the years, many regulatory bodies of the state and also the construction industry have been passionate about sustainability and create some regulatory regimes and rating systems for establishing construction quality benchmarks in terms of sustainability. One of the first and foremost proposed is the Code of ethics which was adopted by the American Society of Civil Engineers (ASCE) in 1914, addressing the responsibility of engineers toward society [1]. The initial seven proposed canons never talked about nature and built environment but in 1983 the 8th canon was introduced, which talks about the services of engineers toward the environment for the benefit of people and future generations. International Building Performance Simulation Association (IBPSA) also offers procedures related to building performance, construction process and review of building model developments, utilization of software programs and standardization [2]. Another leading organization BREEAM proposed the world's first sustainability assessment, rating and certifying methodology for Buildings, Infrastructure and for master planning projects [3].

The joint organizations of US (USGBC)and Canada Green building councils (CGBC) proposed a Leadership in Energy and Environmental Design (LEED) rating system for buildings in different categories with up to 110 points [4], with categories like Platinum, Gold, Silver and LEED certified. In Green Globes, the rating system assigns score and for a score of 85 to 100% four Green Globes are awarded, if it is 70 to 84% three Green Globes, 55 to 69% two Green Globes and one Green Globe for 35 to 54% in various categories of buildings [5]. A highly comprehensive proposal on sustainable development goals (SDGs) was given by the United Nations in 2015 [6]. Of the total 17 goals essentially goals 9, 11, 12 and 13 pertain to construction activities and its sustainability. National Institute of Standard and Technology developed a software, Building for Environmental and Economic Sustainability (BEES), which analyses the environmental and economic performance of building products in all categories by using life cycle assessment method [7]. Until now many government and non-profit organizations evaluated and assessed sustainability and its performance related to environment and other categories as described in their own desirable strategies.

But no organization ever evaluated the rating of even the most common construction material, to be specific concrete, in terms of its overall performance through an appropriate evaluation system based on quantitative test results. In this regard the effort should be to postulate specific criteria, tests and quantitative assessment strategy (grading levels) for different materials, structural facilities, internal and external building environmental parameters etc., to ensure an accurate and non -subjective evaluation. The attempts in that direction are indeed made in a limited way but will need a much bigger effort to be able to objectively quantify sustainability and its accreditation. Also, the proposed procedure should be dovetailed to be in consonance with the local and regional practices and policies, which is sadly lacking in the present practices. Such an approach could reflect the sustainability of the more exotic materials like UHPC in their proper and appropriate perspective in any specific scenario.

5. Materials for UHPC in Relation to Sustainability

In the grand scheme of the evolution of sustainable cementitious composites in the construction scenario, the primary target is to ensure an appropriate utilization of large quantities of supplementary cementitious materials for producing cements of marginal strengths required for most general-purpose applications (termed as sustainable cements of today). Incidentally the International code on cements talks almost of about 27 different types, which could literally encompass many of these use without much effort. It was felt appropriate to leave it to the individual manufacturers to choose the varieties they wish to produce and market with an appropriate notification of the type and grade of material that is being supplied. Incidentally, one should understand that the grade of cement is a representation of its strength at a water cement ratio 0.50 in almost all the codes world over, but for the Indian standards. This aspect puts the concrete designer at a serious disadvantage, considering that we need to compete in an international scene today. Of course, it is a fact that cement is only an intermediary ingredient in the concrete manufacture, and the strength of concrete as desired by the user with the appropriate factor of safety and workability has to be ensured. At the other end of the spectrum, we have cementitious material compositions that could produce ultra-high performance composites, and their role and utility have to be appropriately understood and adopted.

UHPCs requiring significant quantities of cement as well as SCMs and fillers apart from polycarboxylate based superplasticizers and graded aggregates to achieve a defect free hydrated composite always inhibits any notion of sustainability. However, looking at the possibility of utilizing much smaller quantities of these composites, particularly if adopted judiciously both in terms of size, shape and structural form, it is possible to utilize its excellent strength properties to arrive at lasting and sustainable structures. A critical understanding of the aforementioned concepts and possibilities obviously is only possible after a comprehensive evaluation of the appropriate data available in making UHPCs possible. The literature in recent times is resplendent with several publications claiming to have achieved UHPCs of various hues without an appropriate definition for the composite, though it is possible to arrive at a UHPC with a very wide variety of materials and their combinations. Even so the performance characteristics and their evaluation strategies for applications requiring to perform under varying structural and environmental loads is always a matter of concern. With this in view, a brief outline of the materials associated with the production of UHPCs is discussed hereunder. Figure 2 presents a thematic representation of the production and processing of the ultra-high performance concrete. It also indicates a few of the methodologies adopted in its development but these are not being discussed in detail presently.

5.1. Cementitious materials

The development of UHPC is not only for its ultra-high strength but also to satisfy the requirements of the performance characteristics. Due to the higher binder content, production cost will increase and it will automatically impact the environment in terms of what is generally recognized as the carbon footprint. To overcome these issues related to cement and to have a control on the heat of hydration, most researchers incorporate supplementary cementitious materials Researchers generally used ASTM Type-1 or an OPC 53 Grade cement though there are a few sporadic attempts at using the other types as well. The typical amounts of cement content in UHPC varies widely, varying from around 800 to 1400 kg/m3. Alkaysi achieved a compressive strength of 195 MPa at 28 days with Type-1 white cement [8]. Alsalman used a high binder content of 1363 kg/m3 with 3% of fibers and got 179 MPa strength at 28 days [9].

Primarily studies on UHPC utilized silica fume (SF) in combination with a high strength cements because of its high reactivity and the optimum percentage was in the range of 25 to 35% of the cement. Alkaysi introduced 25% of silica fume and achieved around 191 MPa strength at 28 days [8]. Huang used 30% of SF and obtained the highest strength as 150 MPa. Apart from SF, cement can also be replaced with other mineral admixtures like ground granulated blast furnace slag (GGBS) to improve the fresh and hardened concrete properties [10]. Yalcinkaya replaced cement with 50% GGBS and got 182 MPa strength [11]. Similarly, cement was also replaced by fly ash (FA) which could improve workability of concrete but attains strength at later ages.

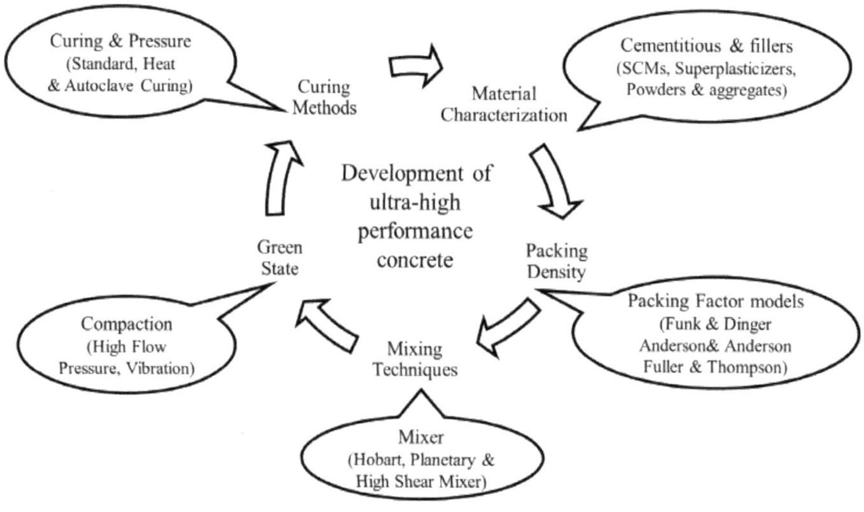

Figure 2. Outline of the production and processing of UHPCs.

5.2. *Superplasticizer*
Superplasticizers are also best known for improving the fluidity and for better coating of the paste to the aggregates. The typical w/c ratio for UHPC concretes is about 0.15 to 0.25 and with this w/c ratio making concrete is obviously difficult. Also, the higher binder content with the associated higher specific surface area of materials will increase the demand of water for wetting. To alleviate these issues associated with water requirement polycarboxylic ether-based superplasticizers are used in UHPCs. Alkaysi used 1.35 % HRWR and got highest strength as 195 MPa [8].

5.3. *Fillers*
To be less expensive than binders, fillers are used and they occupy more or less 70% volume in concrete. In general, UHPC matrix is not having coarse aggregate (CA). However, to compensate, fillers like quartz powder, limestone powder, are adopted. Le Hoang utilized quartz sand and achieved the maximum strength as 198 MPa with the addition of 20% silica fume [12]. Mo used 30% limestone powder in UHPC mixes but obtained lower strengths of only around 90 MPa [13].

5.4. *Coarse aggregates*
It is well accepted that by reducing the coarse aggregate size ensures highest strength concrete. However, in one takes due care to ensure their distribution appropriately, it is still possible to make UHPCs with coarse aggregate sizes ranging up to maybe 10 mm. Though efforts have been sporadic, there are few who clearly were able to overcome the associated issues and produced UHPCs.

5.5. *Fibrous additions*
Fibers are best known for imparting properties like fracture toughness, ductility and resistance to impact in concrete composites. In general, different fibers like steel, carbon, glass and polypropylene have been used in UHPC matrices. Hannawi used different fibers like steel, wollastonite, basalt, PVA, PP-PE in UHPC mixes [14]. The highest compressive strength achieved was 173 MPa for steel fibers and the lowest is for synthetic fibers when compared with control mix. Jiayuan He utilized glass and high-performance polypropylene (HPP) in UHPC at varying percentages [15]. The highest strength for glass fibers and HPP are at 1% and 2% and the combination shows more or less similar results. Meng used carbon nano fibers (CNF) and strength were incremental to percentage of incorporation and highest strength results achieved was more than 180 MPa at 0.3% [16].

6. Mixing, casting and curing details
Mix design is a tool for designing the concrete mixes and needs occasionally modifications according to the locally available raw materials. UHPC matrix with high binder and powder content have to be packed densely and most of the researchers have used the Andreasen and Andreasen model for the optimization of materials and for best packing density. UHPC matrix can be prepared by using several options like different mixing sequences, vacuum and pressure compaction during setting to avoid the pores and achieves excellent mechanical properties. Codal provisions for the design and testing of UHPC were proposed by a few and are basically generic only. The mixing, casting and curing operations vary widely and were seen to be influencing the characteristics of the final product significantly. Generally, it is suggested to mix the binder and powders uniformly for a few minutes then add water and SP with a further mixing for about 5 minutes continuously. The fibers are added when the concrete is showing fluidity. However, curing is done at standard room temperature or heat / steam curing or autoclaving.

7. Performance of UHPC and Sustainability
Strength is the most important criteria in the design of concrete structures, though performance over its life depends upon the environment. In UHPC, strength and performance play a critical role for ensuring not only economy but also sustainability. Strength will be influenced by so many parameters especially binder content, water/binder (w/b) ratio, reactive pozzolanas and fibers and curing regimes, just to name

a few. Due to the high binder content, less w/b ratio some of the cement will be left un-hydrated and it will automatically affect the strength and its enhancement in several ways. So, proper optimal packing of cementitious composites is necessary to understand the behaviour of UHPC. The flexural and tensile strengths of UHPC vary widely depending on the fiber content in most cases.

Yu used three different binders and w/b ratios and got a highest strength of 156 MPa with a high binder content of 874 kg/m3 and lower w/b ratio of 0.23 [17]. Yu also used three types of fibers as long straight (LSF), short straight (SSF) and hooked (HF) and the highest strength results achieved as 142 MPa at LSF. Tanvir Ahmed found that compressive strength, flexure and tensile strength reduced 54%, 56% and 54% due to varying percentages of fly ash and w/b ratio of mixes [18]. Zhiwu Zuo found an increase in compressive and flexural strengths of UHPC at 28 days, 161.5 MPa and 44.9 MPa under steam curing, compared to standard room curing [19]. Šahmenko used short steel fiber (SSF), carbon fiber (CF), and combination for both high strength and fracture energy [20]. Safeer Abbas used different steel fibers lengths, 8, 12 and 16 mm at percentages of 1,3 and 6% and achieved the highest strength of 173 MPa with 3%, 12 mm fiber at 28 days. The modulus of elasticity was 43.4 GPa for 1% SSF with 16 mm fiber at 28 days. The tensile strength was 39.8 at 6% of SSF with 8 mm fiber at 28 days [21].

Durability is simply known as performance which is indicative of the service life of the structures. According to Portland Cement Association, UHPC is a sustainable and durable concrete with 100% Freeze/thaw resistance after 300 cycles, salt-scaling of less than 0.013 mm, abrasion resistance of 1.7 % and oxygen permeability of less than 10-19 ft2. Alkaysi examined the freeze-thaw resistance of UHPC mixes showing incredible results. The rapid chloride permeability charge passing falls under negligible category and it is in the range of 57 to 939 coulombs for varying silica fume percentages [8]. Safeer Abbas determined the volume of permeable voids (VPV) which was less than 4%. The sorptivity coefficients also reduced when compared to HPC. Steel fiber length and dosage, however, doesn't show a significant effect on RCPT or on chloride exposure of UHPC [21]. These abundantly support the fact that the life cycle cost of UHPCs could be lower than even high strength high performance concrete hitherto produced, primarily due to the dense microstructure realized.

8. Appropriate Structural Concepts for Effective Utilization

The usefulness of higher compressive strength in the load bearing column structural members evidently is quite understandable. Also, it can be seen from the literature that there are a few evaluations of the possible size reductions even in flexural members, showing the possibility of almost nearing the corresponding steel sections, through the use of UHPC. Such one-to-one comparisons with structural steel sections cannot do justice to the possibilities offered by UHPCs, as it can be easily modulated to accommodate and alleviate the problems of the buckling of unsupported and extended flanges and webs through suitable design. Also, there have been several architectural shapes and forms which are very difficult with materials or metals like steel, but which can be simply cast to the appropriate recommendations through limited formwork or by prefabrication. It is also noted that for long span bridge girders the advantage of reduced sizes and consequent weight reduction possible is a phenomenal advantage at all times. The fact that there have been several exemplary uses of UHPC for bridge members is an ample proof of such an advantage.

9. Overall Sustainability Effectiveness of UHPC

As already stated, being a highly dense material the ingress of harmful environmental agents that degrade the structural components is literally absent giving the structures a significantly improved life span at all times. A lot can be achieved in the hands of an knowledgeable structural engineer, through properly recognizing the material properties and possibilities of its amenability to be cast into appropriate shapes and forms to cater to the needs of the structure and the environment in which it is supposed to perform.

10. Conclusions

Ultra-high performance concrete is the new generation sustainable concrete with incredible application possibilities in all sorts of structures in construction industry and satisfy the dreams of both architectures and engineers. The present era is mainly interested in sustainability and preserve the environmental resources to the next millennia and for the future generations without compromising their own needs. The following conclusions have been made for the better sustainable UHPC for the future of concrete structures:

- The low w/b ratios and optimal packing of cementitious composites in UHPCs promote significantly higher strength and performance without compromising on other factors.
- The higher binder content can be effectively minimized through eco-friendly reactive pozzolans, improving the microstructure, packing density and homogeneity due to its higher powder content. High temperature curing shows possible early strength gain which can be suitably exploited in the case of prefabricated structures.
- Even though UHPC is expensive due to its high binder content, its enhanced properties of flowability, strength and ductility and durability characteristics make it possible to achieve more economical and lasting constructions requiring minimal maintenance.
- Its strength and performance simply make it highly sustainable for critical infrastructure like bridges and nuclear structures. Apart from the above, UHPCs fulfil the dreams of engineers and architects to literally play with shape and form, achieving sleek and slender structures.

References

[1] Vesilind P A and Alastair S G 1998 Sustainable development and the ASCE code of ethics *Journal of Professional Issues in Engineering Education and Practice* 124(3) pp: 72-74.
[2] Hensen J L M and Clarke J A 1999 Building simulation: state-of-the-art and the role of IBPSA *Proc. Symp. Modellen voor gebouw-en installatiesimulatie* V.19 (Delft) pp:1-9.
[3] Lee W L and Burnett J 2008 Benchmarking energy use assessment of HK-BEAM BREEAM and LEED *Building and environment* Nov 1 43(11) pp: 1882-1891.
[4] Azhar S Carlton W A Olsen D and Ahmad I 2011 Building information modeling for sustainable design and LEED® rating analysis *Automation in construction* 20(2) pp: 217-224.
[5] Say C and Wood A 2008 Sustainable rating systems around the world *Council on Tall Buildings and Urban Habitat Journal (CTBUH Review)* 2 pp: 18-29.
[6] SDG UN 2018 *Sustainable development goals* (United Nations).
[7] Lippiatt B 1998 Building for environmental and economic sustainability (BEES) *Construction and the environment CIB World Congress* (Sweden) June.
[8] Alkaysi M El-Tawil S Liu Z and Hansen W 2016 Effects of silica powder and cement type on durability of ultra high performance concrete (UHPC) *Cement and Concrete Composites* Feb 1 66 pp: 47-56.
[9] Alsalman A Dang C N Martí-Vargas J R and Hale W M 2020 Mixture-proportioning of economical UHPC mixture *Journal of Building Engineering* Jan 1 27 pp: 1-15.
[10] Huang H Gao X Wang H and Ye H 2017 Influence of rice husk ash on strength and permeability of ultra-high performance concrete *Construction and Building Materials* Sep 15 149 pp: 621-628.
[11] Yalçınkaya Ç and Çopuroğlu O 2021 Elephant skin formation on UHPC surface: Effects of climatic condition and blast furnace slag content *Construction and Building Materials* Jan 25 268 pp: 1-12.
[12] Le Hoang A and Fehling E 2017 Influence of steel fiber content and aspect ratio on the uniaxial tensile and compressive behavior of ultra-high performance concrete *Construction and Building Materials* Oct 30 153 pp: 790-806.

[13] Mo Z Wang R and Gao X 2020 Hydration and mechanical properties of UHPC matrix containing limestone and different levels of metakaolin *Construction and Building Materials* Sep 30 256 pp: 1-10.

[14] Hannawi K Bian H Prince-Agbodjan W and Raghavan B 2016 Effect of different types of fibers on the microstructure and the mechanical behavior of ultra-high performance fiber-reinforced concretes *Composites Part B: Engineering* Feb 1 86 pp: 214-220.

[15] He J Chen W Zhang B Yu J and Liu H 2021 The mechanical properties and damage evolution of UHPC reinforced with glass fibers and high-performance polypropylene fibers. *Materials* 14 pp: 1-24.

[16] Meng W and Khayat K H 2016 Mechanical properties of ultra-high performance concrete enhanced with graphite nanoplatelets and carbon nanofibers *Composites Part B: Engineering* Dec 15 107 pp: 113-122.

[17] Yu R Spiesz P H and Brouwers H J 2016 Development of an eco-friendly ultra-high Performance Concrete (UHPC) with efficient cement and mineral admixtures uses *Cement and Concrete Composites* Jan 1 55 pp: 383-394.

[18] Ahmed T Elchalakani M Karrech A Ali M and Guo L 2021 Development of ECO-UHPC with very-low-C3A cement and ground granulated blast-furnace slag *Construction and Building Materials* 284 pp: 1-17.

[19] Zuo Z Zhang J Li B Shen C Xin G and Chen X 2022 Effect of curing regime on the mechanical strength hydration and microstructure of ecological ultra-high performance Concrete (EUHPC) *Materials* Feb 23 15(5) pp: 1-19.

[20] Šahmenko G Krasnikovs A Lukašenoks A and Eiduks M 2015 Ultra-high performance concrete reinforced with short steel and carbon fibers *Environment. Technologies. Resources. Proc. of the Int. Scientific and Practical Conf.* June 16 vol. 1 pp: 193-199.

[21] Abbas S Soliman A M and Nehdi M L 2015 Exploring mechanical and durability properties of ultra-high performance concrete incorporating various steel fiber lengths and dosages *Construction and Building Materials* Jan 30 75 pp: 429-441.

Sustainable strategies for monument protection - A panoramic view of the possibilities and pitfalls

Ganesh Babu Kodeboyina *

* Department of Civil Engineering, Mahindra University, Hyderabad, India.

*Corresponding author: ganesh.kodeboyina@mahindrauniversity.edu.in

Abstract. It is indeed a fact that the rich structural heritage is at the crossroads of extinction either due to the overexploitation of their purported existence for tourism or alternatively due to our abject apathy to their very existence. Apart from this, the prospect of irrevocably losing even the few spattering of monuments that we seem to protect and guard against their degradation through inappropriate, or in some cases highly casual and ill-informed strategies is one aspect that has attracted the attention of the global community in recent years. While, there appears to be an effort to put a system in place to ensure a proper repair and rehabilitation of these monuments, there is still a lack of understanding of the broad principles to ensure an effective yet sustainable maintenance and protection strategy. To start with these ancient monuments can be broadly characterized into few specific categories based on the materials used in their construction. The paper attempts to delineate some of them and look at the tenets to be followed to ensure a sustainable upkeep of these monuments. In doing so, a few examples of the type of problems associated with monuments in the Indian context were also looked into as illustrations.
Keywords: Sustainability, Monuments Construction Materials, Rehabilitation, Protection, Maintenance.

1. Introduction

Monuments give us an opportunity to have glimpse into the history of mankind and many a time their very existence leaves us in awe, as we realize that with even the present-day advancements in technology and science, we may still not be able to replicate some of them even in part. They also serve the specific purpose of being a window into our past, the vanished and forgotten civilizations, the lost technologies and their uncanny ability to do things which require a phenomenal scientific understanding and thought. The fact that several civilizations located on either side of the globe, even without any apparent or possible connections, have been able to predict the calendars, seasons and stellar motions with uncanny accuracy, have their own mathematical rules for them like in the Mayan, Indian and other cultures is only a fact to be astounded and admired. It is imperative to realize at the very start of this particular paper that even a simple overview of the kind of intricacies involved that are there are almost impossible to enumerate. Needless to say, that on several occasions one has not even an inkling as to why they came into being, who constructed them and for what purpose. The few lines above clearly suggest that not every old construction can simply be adjudged as an important candidate for being termed as a monument. Naturally monuments are classified into different levels of importance, world heritage sites, national monuments, local memorials apart from religious shrines, all with their different levels of significance, only to be perceived individually or by a community. The different parameters associated with such classification is difficult to explain and essentially depends on their age, aesthetics, architecture, skill apart from technical brilliance, all or some of which form the basis. It is only appropriate to say that many a time the choice of the level is more or less by a certain popularity and promotion, like in all other walks of life, and an application or submission for recognition to the authorities concerned. Inevitably, there are certainly some that get left out in spite of their actual significance and value. But of course, there has to be a system, however imperfect it is. Maybe with time the system will correct itself to find the right answers and an appropriate recognition of the others that also richly deserve recognition and upkeep.

In this context, it may probably be not out of place to mention that there are such entities that have no human hand in their creation, and have come into existence because of the forces of nature, whose geological features or vegetation and forestry or even a natural formation is captivating, if not awe-inspiring. It is only appropriate to say that these geological formations or forestlands, which are indeed designated as national parks by a government be safeguarded and given the same amount of importance in preserving for the generations to come. The possibility of mining, development of even tourism could tarnish the pristine originality of these locations and should be undertaken with the necessary precautions and care. However, having taken a pretty broad global viewpoint of what needs to be preserved as monuments, it is important to realize that attempting a discussion at such all-encompassing levels is not possible. The present discussions are limited to only constructed facilities which have a broad significance for understanding the transformations in human civilization over time, be it in the recent past or the traces of assets that depict the bygone civilizations.

Naturally there are several publications and reports that are specific to a few well-recognized monuments and their preservation efforts. It may be important to realize that in their anxiety or the immediate necessity to preserve to the extent possible some of these axioms of monument protection like the unacceptability of local or global intrusive efforts or use of foreign and different materials are often adopted. It may also be only because the fact that they do not appreciate or specifically understand ancient texts and tenets that these structural and material configurations were based on. A typical example of such ancient, well-choreographed texts are the several treatises on the construction of temples and shrines in India [1 – 5]. In fact, these texts or shastras as they are called have been handed over for generations from times immemorial, but in some cases were also codified into texts at some point in the unknown past. These axioms could be used not just for the creation of buildings and shrines alone but could be adopted for setting up complete towns and forts and huge metropolises like in the case of Harappa and Mohenjo-Daro civilizations. Many a time the scientific significance is lost in transmission and execution. There are several other such examples all around the globe, be it the Roman, or even Mayan and Inca civilizations. One has to have a healthy respect for the history and writings of that period to appreciate and understand the significance, philosophy, structure and material used in the conceptualization and construction of these monuments. It is only then that justice can be done to the protection, repair and rehabilitation of these ancient artifacts that we should be preserving for the future generations in their original form and sprit.

The idea of this paper is not to replicate the different aspects of the protection of monuments or of any specific monument, but to look at it from outside an establish broad and all-pervasive guidelines for an effective implementation of the process in a scientific and sustainable regime. In fact, the discussions on the defects and distress in monuments and the proposed possibilities while outlining the pitfalls in the protection strategies adopted are based on the personal experiences during the visits to many national monuments and world heritage sites visited as a responsible and active participant in the rehabilitation efforts. It is also necessary to point out that apart from the experiences in field, a lot of background information gathered and studied, both the scriptural and modern texts their commentaries and explanations as well as the practices of various countries, formed the basis for these discussions.

2. Characterization of monuments

It is a very difficult proposition to look at the broad classification of the enormous number of monuments in any country, given the fact that apart from the remnants of civilizations dating back to the prehistoric era that are not so obvious in most cases there are several others which represent a more recent history in terms of the religious shrines, forts, palaces during the reigns by the different rulers of that period. In a country like India alone the abundance and phenomenal complexity of the monuments and their ancient history involving the renovation, rebuilding and expansions by several rulers over this vast period makes it impossible to catalog and keep up to date the requirements, even for a dedicated governmental organization like the "Archaeological survey of India" (ASI). Also, with the priorities of the government in ensuring the well-being of its people (the requirements of food, shelter, health etc.) far outweighing any other requirements, efforts are made to strike a balance in ensuring that at least they protect the most prominent historical structures to the extent possible with the limited resources.

Naturally, the task of establishing even a basic yet comprehensive database of the enormous variety of monuments, memorials and shrines that need to be protected and to keep track of their specific needs from time to time is probably well beyond the ability of any single organization. It would be prudent to setup a pyramidal system in which the apex body, with the authority of controlling or at least monitoring the provincial or state, district or regional and local bodies, be constituted for the purpose of maintaining the heritage structures. Also, there are several philanthropic and business organizations who can chip in and help in the task above, with the understanding that they are also equal partners in maintaining the structures and preserving historical icons.

Protection of monuments invariably involves expertise from a broad spectrum of disciplines, not only architecture but also the science, technology and engineering of the structure. In particular while dealing with the restoration of a monument the primary philosophy is to bring it fully to its original status in terms of the material, structural and architectural splendor as was planned and enshrined. A broad outline philosophy for ensuring an effective and sustainable solution through the three arms of architectural, structural and material domains is illustrated in Figure 1, which is also very similar to the ever so often repeated concept of the three pillars of sustainability. This sustainability approach through an understanding of the Structural, architectural and Material principles could be termed as the "SAM interactive domain" of management. The interaction between each two arms of these primary concepts is also discussed in detail in the later parts of this particular paper. Thus, it is only prudent to establish a comprehensive strategy and laydown broad guidelines depending on the architectural, structural and material composition of these monuments. This is not necessarily universal and there could also be other ways and methodologies for delineating appropriately the strategy for any particular monument or group of monuments. In the present context the author wishes to present a particular strategy that deals with the primary aspects of aesthetic and architectural exactness, structural strength and stability and finally the material compositions and degradation associated with the structural members and their assemblage. The idea is to probably give an outline that could be fairly enshrine the fundamental principles above in attempting the rehabilitation of most monumental structures.

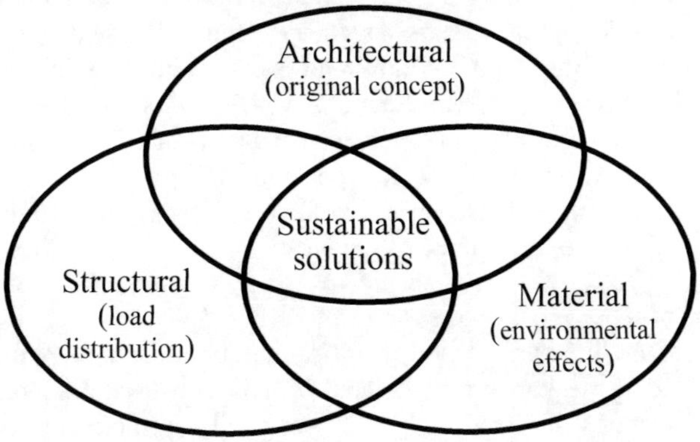

Figure 1. Effective sustainable solution methodology

The author had the privilege of being associated with several investigations dealing with the upkeep of many prominent heritage structures apart from being the consultant for the repair and rehabilitation of many other modern constructions like buildings, bridges, industrial facilities during the past few decades. Some of the aspects regarding the telltale signs, the distress signals, cracks and settlements as well as other aspects that need a careful study cannot even be specifically pointed out both for reasons of their proprietary nature and the impossibility of task in such a short report. One can have a glimpse of some of these aspects, though not in depth as they need to be discussed and published in specific reports from time to time by the various authorities responsible for these structures [6]. Needless to say, that this association with both heritage structures as well as modern constructions facilitated an understanding of how each of them are different. Aesthetics and architecture apart, the need to understand the structural strength and load transfer mechanism for ensuring the stability of the structure is one of the first engineering principles to be recognized [7]. The use of modern construction materials whose characteristics are well understood and were systematically analyzed can only be at best a broad guideline for a better understanding of the materials, methods and other construction aspects of heritage structures. It is with this background that few of the principles that are reasonable to be of consequence in attempting to put up a strategy for the rehabilitation monuments is being attempted herein.

3. Architectural features

It is important to realize that many of these structures, from the prehistoric times to the present day have had their intricacies and maybe even eccentricities that need an appropriate understanding and zeal to preserve the ethos of that particular community and their way of living and thinking. Some of these aspects probably are only enshrined or alluded to in the architectural features and it is essential to preserve the delicate balance of repair, renovation and rehabilitation with the original concept and thought of those who built the structure, many a time with entirely different concepts, materials and with limited resources and skills. The idea is to preserve the originality of those people of the yesteryears who conceived and built these monuments. It is not an easy task to follow the original thought or the scientific principles they are based on. The original technology, skill, knowledge and other resources are obviously not so easy to understand and the architect and engineer in charge should be able to capture the vision and purpose in the creation of the monument. An appropriate resource for understanding these intricacies are probably the texts and treatises of that era or the writings and inscriptions from those periods if available. Even so, transcribing or effectively interpreting these texts is difficult and requires specialists well-versed in that particular domain.

3.1. Architecture – structure interrelations

One of the primary features that is to be recognized in the constructions of the yesteryear is that while floors, walls could be easily erected as they have a firm ground support, the spanning of the roof over the walls needed mostly an arch or a dome. Later years saw the development of wooden rafters spanning these walls with different concepts of making the floors and roofs. Buttresses in various forms and double wall buttress concepts have been adopted for greater wall heights and also to accommodate higher load carrying capacity as the number of floors increased. These adaptations also allowed that the wall thicknesses otherwise demanded for stability could be reduced significantly. Interventions and modifications without the due recognition of these structural factors will only lead to the altering of the character and may result in a local failure or even a total collapse.

3.2. Architectural aspects of sustainability

Primarily, the refurbishment of architectural features is intimately associated with the structural configurations and the load-balancing mechanisms associated with the construction. It requires a definite understanding of the structural features, the load paths, the weak links in the chain to ensure that the rehabilitation is attempted without compromising the stability the structure during repair. As an example, one can talk about the removal of sand support from inside the "Jagmohana" of the Konark Sun Temple after the British filled it for supporting the roof structure almost a century back. The internal damage to the roof structure due to the falling stone of the main temple is well known. But the effects of ageing of these damaged stones in the wet and sand filled environment is to be established for any suitable approach of the sand removal operation. Similarly, during the rebuilding of the Katarmal Sun Temple structure near Almora one has to be cognizant of the original stone used in the various parts. It is indeed a daunting task to ensure the appropriate materials and construction methods on a steep hill slope and in the remote hills of Kumaon. With the eroded stones that were displaced probably due to an earthquake in the past, the site presents an aery ethereal feeling in the chilly atmosphere even on a good summer morning or evening. Needless to say, that the repair and rehabilitation strategy should be planned with due consideration to the skills available locally, supplemented by a limited level of guidance and supervision. It is only appropriate that we recognize the zeal and the spirit of the personnel in organizations like ASI who preserve these few last remnants of yesteryears for the future generations.

4. Structural characteristics

Unlike the aforementioned architectural features and the dichotomy in being able to preserve or not the original thoughts and concept of the bygone era, structural characteristics can be more easily established and can be appropriately analyzed. The main idea of renovation and rehabilitation of monuments should be to ensure that the creator's concept and construction philosophy as well as the technology as enshrined is preserved. It is important to realize that the degradation of the materials and members make it impossible to ensure this. One of the fundamental principles of structural rehabilitation is the principle of adopting non-invasive or non-intrusive methods in the rehabilitation. If ever the replacement of members or replenishment of materials are attempted, it is imperative that one looks for an appropriate replica both in terms of physical and chemical characteristics, to ensure the originality of the structure.

Primarily it is important to understand the associated load patterns on the structure and the load path, the effectiveness of load bearing members and their distress levels in specific. Many a time there is a natural readjustment of the stresses and strains which will lead to minor signs of distress that are to be expected and tolerated. The tolerance limits for such an occurrence in the structure should be such that it does not create any immediate local or global failure. Sometimes these could be unsightly and mar the aesthetics and features of the built structure or even a mural simply. Also, the progressive nature of the distress over time, established in terms of deflections, crack widths and other signs of distress need a careful evaluation. In fact, the concept of not in any way altering the architectural ethos with due regard to the structural configurations and load paths is an essential prerequisite and should be recognized and accepted. Any modulations or modifications without regard to these concepts will certainly lead to unforeseen as well as an unacceptable behaviour of the structure.

Another very important aspect that needs a specific mention in the rehabilitation and protection domain is to have a clarity regarding the progress of deterioration and distress. In fact, it is more important to understand the continuance of the activity of distress once a certain specific defect is noticed. As an example, one could have observed a crack in a particular structure which could probably be due to several causes like settlement, loading effects, thermal effects, shrinkage and environmental changes just to name a few. It is important to know if this particular crack and for that matter any other type of distress is still phenomenon that is in progress or has come to a certain level of stability due to the readjustment of the forces. There are several different methodologies for ensuring the active or passive status of a crack on any other defect - starting from the periodic observation of its development or otherwise, both in terms of width and length of a crack or the inclination, rotation or deflection in structural members. Some of the instrumentation for monitoring progression of defects like cracks could range from simple paper strips, gypsum patties and glass strips to the most advanced displacement transducers like LVDTs. In case of possible vibration scenarios, the use of sensors like accelerometers and data acquisition systems are also adopted, depending upon the importance and cost benefit ratio.

Such problems have been observed in several monumental structures particularly at the corners and joints, or even in domes. It is not only essential to maintain a detailed record of each of these distress types and locations, but also have a periodic monitoring to ensure the continuance. It is also important to assess the possibility of the local or a total failure due to such defects each time and take appropriate measures. In particular, the distress signals or even cracks in walls, domes and roofs of some of the very important monuments or world heritage sites like several mosques and even the Taj Mahal seem to be in neglect and are not adequately monitored. Some of these places are inaccessible even for monitoring. Many a time these hairline cracks are the only signs of distress and degradation due to several factors. At these distress locations establishing the route cause may not be always possible without an understanding of several factors discussed earlier. One simple and effective method to understand distress is to observe the moisture migration path in the structure, which is a excellent and reliable method for understanding distress, but to only a trained eye.

Another aspect which is of importance is to understand the foundation characteristics and the settlements in the foundation due to moisture variations seasonally and over the life of the structure. Some of these changes could have affected the settlement or swelling of the soil below, which could affect the structural performance not just in the near vicinity but relatively far away along the load path, probably due to a possible weakness at that particular location or the inherent lower strength of the location. In a typical wall of a religious shrine which was in distress endangering the nearby minaret a specific method of underpinning of the foundation to ensure its stability was successfully utilized. In fact, the possibility of using a buttress or supporting with some adjacent structure that has the capability to transfer the load appropriately could also be a solution. These solutions are not always unique though the problem associated with the defect as can be seen will be mostly unique. It is for the engineer and architect in charge of the site to look for the most appropriate of the many possible engineering solutions.

4.1. Structure – material interrelation
Any structure is made-up of an array of materials all of which contribute to its overall functioning and will also degrade with time in different ways each at a different rate. While the most important aspect is to ensure an appropriate functioning of the primary loadbearing members and its constant materials for stability, the functioning of non-loadbearing members like fixtures and claddings could have a more important role because of the aesthetics. It is also important to understand that the structural construction methodology could have adopted different layers of materials, like the core and cladding materials on either side apart from the bonding layer in between to ensure an appropriate rehabilitation. There is often a failure related to both the core and cladding due to moisture ingress through the porous and weak bonding layers as observed several monuments. One can observe even with naked eye some such defects in the cladding layers of Qutab Minar. The recognition of the primary source of ingress and the flow path of this moisture is difficult to establish and need a careful observation and a keen eye for detail in terms of the tell-tale signs of its distribution.

4.2. Structural aspects of sustainability
In situations that show a distress in structures, could be in the main wall, buttress or even the roof of the structure, it is essential to understand the reasons for distress and where it originated and not just where it is located. Incidentally, many a time the repair is attempted to look at probably a crack and assume that the part needs strengthening. Any repair or strengthening the location will only force the distress to shift to another location where the stresses are shifting now. Such ad-hoc or unplanned or spur of the moment repairs without giving sufficient thought to its effectiveness in a given situation – like simply filling, bonding or stitching is totally unsustainable. The progressive nature of the distress over time, if it can be established in terms of deflections, cracks and other signs of distress need a careful evaluation. In fact, a concept of not in any way altering the architectural concept with due regard to the structural configurations and load paths is an essential prerequisite and should be recognized and accepted. Any modulations of modifications without regard to these concepts will certainly lead to unforeseen as well as an unacceptable behaviour of the structure.

5. Material characteristics
It is obvious that while the structural form, shape and the member configurations are all dictated by the architecture and the monument as it exists, the stability and load carrying capacity of the same is always defined by the strength and characteristics of the materials of construction. Apart from this the performance and or deterioration of the materials adopted in the construction of the monument will

invariably be defining the strength, effectiveness as well as its durability over time. It is also a fact that the exact composition and production process of the materials of yesteryears is lost in history and those documentations if available are subject to the interpretations of a few many a time with limited knowledge. In fact, even the making of pozzolanic cementitious mortars used by the Romans in Italy is not fully understood and is an active topic of international research in many prestigious universities all around the globe presently. Needless to say, that in the Indian context many of the formulations of the binding materials have always been an enigma and the scriptures that detailed some of them are probably lost forever. In view of this the replenishment of any specific material in a structural member is obviously a difficult proposition even if it was sought to be attempted, as should be the guiding principle.

With this in view the first principle in the characterization of these could be to talk about their materials of construction - like stone, bricks, tiles, lime mortar and lime concrete etc. In fact, the enormous varieties of stone and their degradation due to the environmental forces in any specific structure is in itself highly complex and maybe sometimes subject for investigation. A fine example of this could be the leaching out of the chemical constituents from the khondalite stone, a metamorphic rock used in the construction of the sun temple at Konark in Odisha. The breaking and falling of the peda stone projections due to their own weight in Konark is an excellent example. The ASI has undertaken a special program to understand the degradation and durability, as well as the leachates from such stones, to have a grip on the methodology of restoring monuments using these materials. The work need not be limited to only stones but should be extended to bricks, tiles and mortars used in such structures. Also, there are iron keys as interlocking fixtures to ensure the integrity of certain stone monuments, the sun temple at Konark being one such. While we do talk about the Iron pillar at Qutab minar in Delhi which has not shown any signs of rusting over all these ages, the same is not the case with those iron keys at Konark. Normally, in most stone or even brick masonry monuments, the use of iron fixtures are not attempted for this reason.

5.1. Material – structure interrelations

The outstanding performance of the materials used in the bygone era is only to be viewed with awe as reflected by the minimal deterioration and distress in these monuments. Firstly, it emanates from the fact that most of the structures lasting millennia are primarily Stone monuments. However, there are others like the Pantheon in Rome that lasted almost 1000 years which is still functional. The source of the material at each level of the particular structure, it's composition and the architecture associated with it has made this possible, though one should appreciate the fact that it is the one structure that is maintained in pristine condition all through the ages for its obvious religious significance. Even so, it is probably pretty difficult to even emulate or simulate the materials even with the present-day knowledge.

5.2. Material aspects of sustainability

It is often said that any repair should be an exact replica of that that existed originally, be it stone, brick, mortar or other cementing material. This is often easier said than done, as it requires not just the composition of the original material matrix which is really difficult to establish. One aspect is that several of the earlier compositions have not been documented and are probably passed on to the next generation only as a family traditional knowledge. The second aspect that complicates the production earlier materials is due to fact that many a time the constituents contain ingredients that are only known to a select few and often this knowledge is lost over time. Also, it may be difficult to obtain such traditional materials as the processing and production of these materials is not known today. Even the few that are in use for the specific purpose of fixing an idol to the pedestal in the temples is presently done by only a select few who have been making similar materials for generations. While they have guarded these as hereditary treasures by those few who have access to them from their forefathers, the authenticity or even the background axioms from the shastras on which they are based is at best unclear. The phenomenal number of monuments that are existing in this country from the different periods and reigns of different rulers who used entirely different philosophies in the constructions of their time along with the local and regional modifications associated with them make it almost impossible to put a fix on the type of repair material to be used today. Even the present day scientific and analytical tools in the most advanced laboratories will still not be able to give a definite answer to many of those materials and of their ingredients and production processes. This only accentuates the problem of compatibility between the existing materials of construction and the other replenishments that are associated with the rehabilitation of monuments.

At this stage it is only appropriate to point out that the discussions presented in this paper, in terms of the architectural, structural and material characteristics are not the only aspects that need a specific understanding. Also, it should be understood that all the three have an equal significance in ensuring a holistic, lasting and sustainable repair and rehabilitation strategy. However, maybe it is not out of place to mention that the weakest link in the rehabilitation strategies adopted to date not only in the Indian context but world over is the fact that there is not enough thrust towards an understanding of the material characteristics and its degradation, particularly in relation to the environment as well as its percolation through the structure. The reasons are probably obvious as can be seen by the fact that the final expected result reflected in preserving its original architecture and beauty is not fully realised. Incidentally, it is also necessary that the monument should be structurally stable necessitating an elaborate structural framework and scaffoldings as you often see during the rehabilitation efforts. One can only presume and hope that dew cognizance to the material degradation parameters and its effects on the structural strength and stability are understood appropriately. Also, our anxiety to ensure a reasonable level of renovation in the shortest possible time with least physical and financial efforts. The need to dwell deep into the original material constituents, their processing and application as well as their degradation over the centuries is always a painstaking and time-consuming effort that falls on the wayside. Most rehabilitations look for immediate solutions, however inappropriate and unimaginative they may be. It is thus the primary task of this paper to present before the authorities concerned a holistic picture while pointing out that there are several other aspects that need an understanding for a satisfactory, holistic and sustainable monument protection and rehabilitation.

6. Concluding remarks

A broad perspective in terms of the architectural, structural and material parameters associated with the protection and rehabilitation monuments was attempted. Some of the critical aspects broadly relating to the possible pitfalls in the assessment methodologies and concepts to address such possibilities have been discussed. One notable aspect that has not been discussed or taken up is about the test procedures necessary for understanding the distress levels, which are relegated to a future discussion. While it is obvious that the three parameters, architectural, structural and material, and all significant in their own right, the present status of not getting enough thrust towards an understanding of the degradation of materials is clearly to be stressed.

References

[1] Dagens B 1994 Mayamatam Vols. I & II. *Indira Gandhi National Centre for the Arts & Motilal Banarasidass Publishers Pvt. Ltd.* India.
[2] Acharya P K 1995. Manasara Series: Vol.1-7 *Oxford University Press* India.
[3] Sharma S K 2012 Samaranagana sutradhara of Bhoja deva *Piramal publications* India p: 950.
[4] Brown P 1942 Indian Architecture: Buddhists and Hindu Period *Taraporevala & Sons* India.
[5] Fergusson J 1910 History of Indian and Eastern Architecture Vol I & II *Low Price Publications* India p: 449 & p: 521.
[6] CAG report 2013 Performance Audit of Preservation and Conservation of Monuments and Antiquities *Ministry of Culture Government of India* Report No 18 p: 300.
[7] Mustafaraj E 2013 Structural assessment of historical buildings: a case study of five Ottoman Mosques in Albania *2nd International Balkans Conference on Challenges of Civil Engineering* p: 13.

Numerical simulation of a floating-type point absorber wave energy converter

Sharath Srinivasamurthy*

* Institute of Ocean Energy, Saga University, 1 Honjo-machi, Saga 840-8502, Japan.

*Corresponding author: sharath@ioes.saga-u.ac.jp

Abstract. Wave energy converters (WEC) have been researched extensively to harvest large amounts of untapped ocean renewable energy. Floating-type point absorbers are a type of WEC that converts the ocean wave energy to electrical energy, by utilizing the relative motion of multi-body system mainly consisting of a floating section and a reaction section. The design of floating-type point absorbers is not straightforward and needs special attention to understand the dynamics of the multi-body system, and power take-off (PTO) mechanism placed in the reaction section. The purpose of this study is to develop a numerical simulation tool which can help understand various dynamics of a floating-type WEC and aid in the practical design of PTO systems. The PTO mechanism is represented using a spring-damper system and theoretical modelling of mean absorbed power is presented. The numerical simulation is developed in Orcaflex and is carried out on a demonstration model to show the usefulness of the developed tool. The heave displacement shows the effect of damping (PTO) on the motion response of the floating-type WEC. It is found that the maximization of absorbed power is possible using the developed simulation tool for a specific sea state. Furthermore, the results show the potential use of the developed tool for other floating-type WEC concepts.
Keywords: Renewable energy, Floating-type wave energy converter (WEC), Mean absorbed power, Representative power take-off (PTO) mechanism.

1. Introduction

Renewable energy has been attracting attention all over the world to reduce carbon emissions and to provide clean energy. Major sources of renewable energy include solar, wind, hydro, geothermal, and ocean renewable energy. Oceans cover about 70% of the earth's surface and has a huge potential to contribute to the energy mix. Ocean renewable energy can be broadly classified into OTEC (Ocean Thermal Energy Conversion), ocean wave energy, ocean current energy, and ocean wind energy. Even though ocean waves around the world are known to hold large amounts of energy, wave energy harvesters are still not as mature as other forms of renewable energy harvesting devices. This limitation is due to the complexities involved in converting wave energy to electricity. In this paper, a type of ocean wave energy converter (WEC) called as point absorbers will be focused since it is an upcoming research field and has good potential for commercialization [1,2].

A point absorber converts the wave energy by utilizing the heave motion (up and down motion) (see Figure 1). Typically, it is either a single-body that generates energy by reacting against a fixed seabed frame, or it is a multiple-body device that generates energy from the relative motion between the two floating bodies. In the case of floating-type WEC system, it generally uses a two-body heaving system, which converts energy from the relative motion between the two oscillating bodies, a float section, and a reaction section [3]. The float is only allowed to move freely in one degree of freedom, with respect to the reaction section (generally heave motion). To understand point absorber behaviour, precise modelling tools are needed to simulate the dynamics of the device, along with a well-defined combination of modelling methods. Technology developers, engineers, and researchers need a tool which employs the relative motion of the multi-body dynamics. The purpose of this study is to develop a numerical simulation tool which aids in the study of floating-type point absorbers. Specifically, a simulation method to maximize mechanical power under a particular sea state is demonstrated by use of a damper system to represent the power take-off (PTO) system.

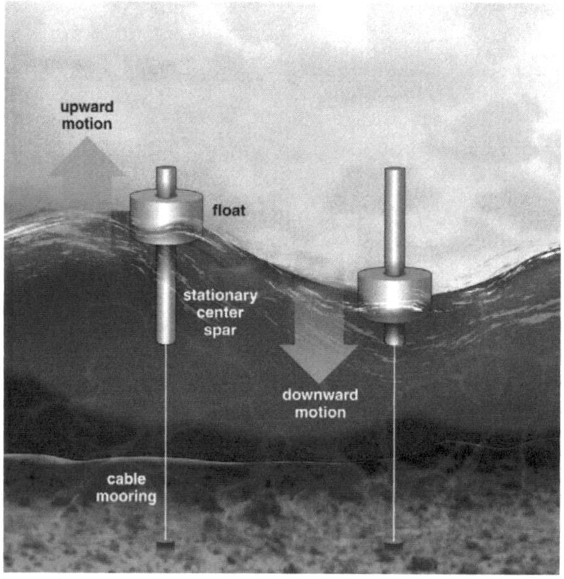

Figure 1. Illustration of a point absorber wave energy converter (Adapted from [3]).

1.1. Modelling of floating-type wave energy converter

Orcaflex is used for modelling the floating-type wave energy converter. Orcaflex is a dynamics analysis software for offshore marine system operations [4]. It consists of various objects to model different components of the device. In this study, both the spar and float of the WEC device are modelled as 6D buoy objects (see Figure 2).

The spar is the reaction section, which is stabilized with the use of 3 catenary-type mooring lines. A link object (similar to a spring-damper system) is used to represent a power take-off (PTO) system. As explained earlier, since there is relative motion between the spar and float in a floating-type point absorber WEC, it is not a straightforward process to set up the conditions of simulation with simple objects. Therefore, in this study, we adopted the use of constraint object in Orcaflex. Constraint is one of the versatile connection objects which can be defined in Orcaflex. Two constraints are defined, one each for the float and the spar sections. The float section is constrained to motion only in heave direction, and connected to the link object. Therefore, when the float section heaves due to the incoming wave, we can control the motion response by varying the stiffness and damping conditions of the link object. In other words, the representative PTO can be modelled which is highly useful for wave energy harvesting devices.

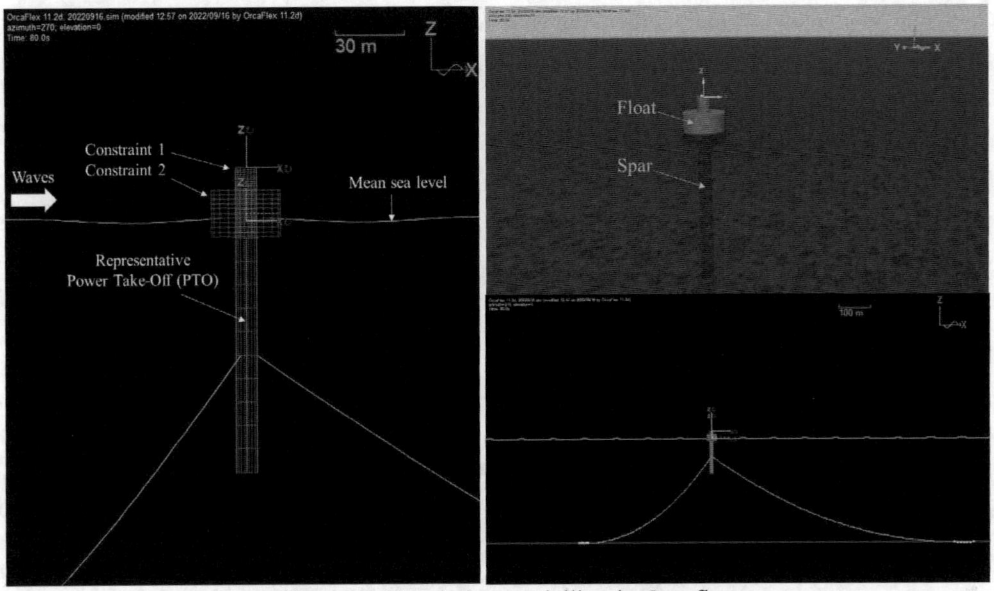

Figure 2. Simulation modelling in Orcaflex.

1.2. Model specifications
This section describes the demonstration model of the floating-type point absorber device. Table 1 shows the principal particulars of the demonstration model. It is noted that the model defined here is hypothetical and used only to demonstrate the development of the simulation tool.

Table 1. Principal particulars of the demonstration model.

Specifications	Value	Unit
Diameter of spar section	9.4	m
Diameter of float section	30.0	m
Length of spar section	130.0	m
Length of float section	20.0	m
Number of mooring lines	3	nos

1.3. Theoretical modelling of mean absorbed power
The mean absorbed power [5] per wave cycle of the moving float can be obtained as shown in the following Equation (1).

$$P_m = 0.5D\omega^2(\tilde{x})^2 \tag{1}$$

Here, D: Damping of the PTO system, ω: Wave angular frequency, $\tilde{x}$: Heave displacement. Effectively, it describes the mean force acting on the float multiplied by the velocity of the float. This simple theoretical modelling will help understand how heave motion can be converted into power. It can be observed that mean absorbed power per each wave cycle is proportional to the PTO damping. Also, it is proportional to square of the heave displacement.

1.4. Simulation conditions
In the demonstration simulation, the mechanical stiffness is kept constant and damping is varied to represent the effect of PTO mechanism. Wave period is set to 8 seconds, and wave height is set to 2

meters in the simulation. Table 2 shows the simulation conditions. Overall, 6 cases of damping are considered and compared with the base case (Case0) without damping.

Table 2. Simulation conditions.

Simulation cases	Stiffness (kN/m)	Damping (kNs/m)
Case1	500	2,500
Case2	500	5,000
Case3	500	7,500
Case4	500	15,000
Case5	500	20,000
Case6	500	25,000

2. Results and discussion

Figure 3 shows the simulation result comparing the effect of damping on the heave motion of float. In the Orcaflex simulation, statics is firstly solved followed by the dynamics. The simulation is divided into 2 stages; Stage 1 is from -8 to 0 seconds, and the wave conditions are set using a ramp function in this duration. Stage 2 represents the float response to incoming wave (0 to 80 seconds). The wave period of the simulation is 8 seconds, and therefore we can observe 10 peaks between 0 and 80 seconds. The blue solid line shows the response of float when damping value is set to zero and a heave displacement of 0.8 meters can be observed. However, when the damping value of 25,000 kNs/m (Case6) is simulated, the heave displacement (dotted purple line) reduces to about 0.2 meters. This shows the effect of damping, and represent the PTO mechanism. In other words, the PTO in reality dampens the WEC device and reduces the motion response while generating available power. So, to generate maximum power, evaluation based on varying damping is essential.

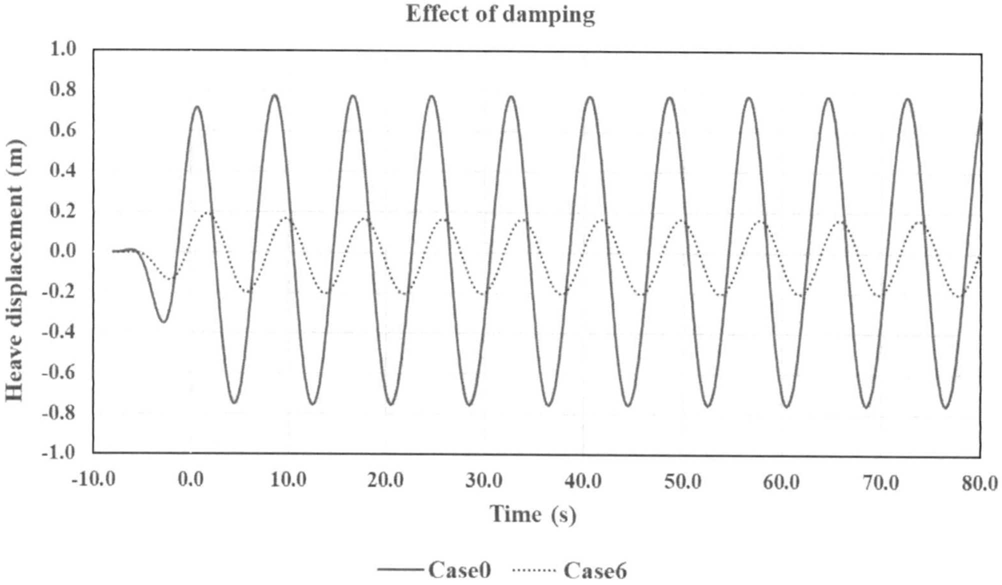

Figure 3. Effect of damping on the heave motion response of float section – Case0 vs Case6.

Figure 4 shows the simulation result of the heave displacement for different damping conditions (from Case1 to Case6). It can be observed that the response of heave displacement decreases as the damping value becomes larger. It can be said that selection of optimum PTO damping needs attention and it can help to maximum power.

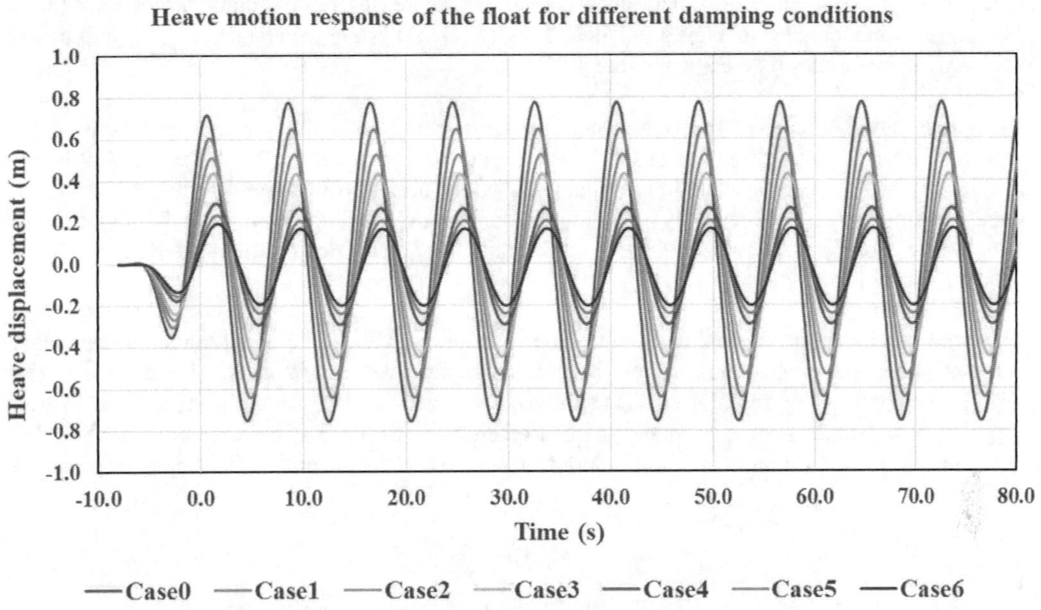

Figure 4. Heave motion response for different damping conditions.

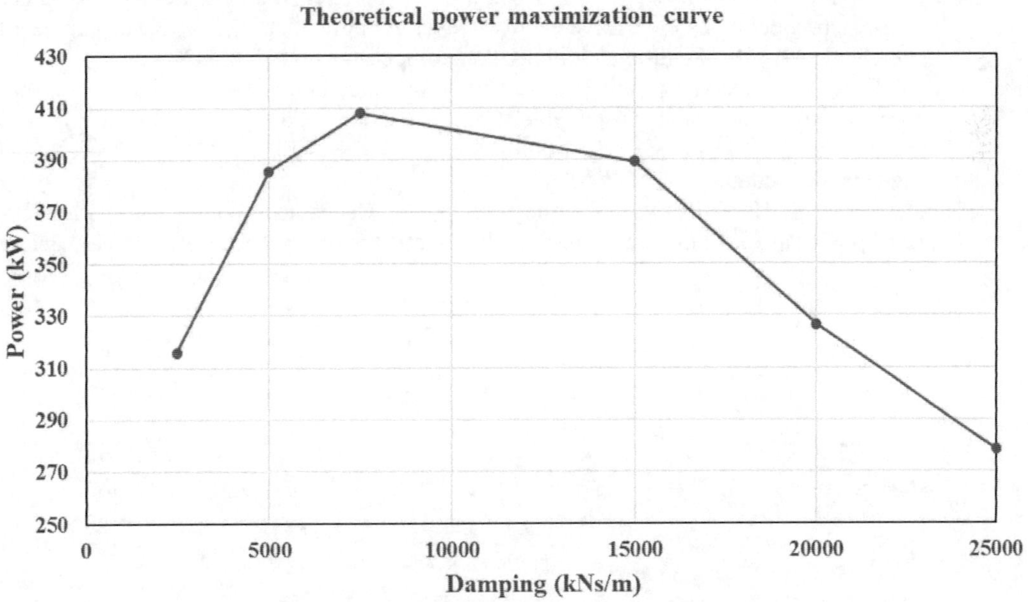

Figure 5. Theoretical power maximization curve.

As discussed in Section 1.3, the mean absorbed power is proportional to square of heave displacement and also proportional to damping value. To understand this, mean absorbed power is plotted for all the

6 cases of damping as shown in Figure 5. The larger damping value need not always produce maximum energy since it also decreases the heave motion and Figure 5 confirms the same. From the theoretical power maximization curve, it is found that maximum power of 410kW can be obtained if the damping value is 7,500 kNs/m (Case3). The absorbed power decreases for damping larger than this value. Therefore, in the real device it is a good idea to select a PTO system which produces a damping of 7,500 kNs/m by using appropriate control strategies.

Using this simulation tool, the power maximization can be studied for different wave conditions depending on the installation site of the floating-type WEC device. Furthermore, PTO system can be developed for the specific installation site. This study demonstrated the advantages of the tool for understanding the float response by including the damping value of the system. In the future it is desired to carefully consider the device specifications and improve the performance of the float.

3. Conclusions

This study focused on developing a simulation tool for evaluating the performance of a floating-type point absorber wave energy converter. The tool demonstrated how to maximize the mean absorbed power by including a representative PTO and heave motion of the float section. It also demonstrated the use of spring-damper system to model the characteristics of a PTO system for a wave condition. The potential use of the simulation method is demonstrated for other concepts of floating-type point absorber WEC.

References

[1] Shami AL, Zhang R, Wang X. Point Absorber Wave Energy Harvesters: A Review of Recent Developments. Energies 2019; 12(1), 47. https://doi.org/10.3390/en12010047.
[2] Beatty SJ, Hall M, Buckham BJ, Wild P, Bocking B. Experimental and numerical comparisons of self-reacting point absorber wave energy converters in regular waves. Ocean Engineering 2015; 104(2015) 370–386. https://doi.org/10.1016/j.oceaneng.2015.05.027.
[3] Li Y, Yu YH. A synthesis of numerical methods for modeling wave energy converter-point absorbers. Renewable and Sustainable Energy Reviews 2012; 16 (2012) 4352–4364. https://doi.org/10.1016/j.rser.2011.11.008.
[4] https://www.orcina.com/.
[5] Chandrasekaran S, Harender. Power Generation Using Mechanical Wave Energy Converter. Journal of Ocean and Climate Systems 2012; 3-1. https://doi.org/10.1260/1759-3131.3.1.57.

Numerical Investigation the Effects of Cone Diameters on the Flow Pattern and Separation Efficiency in a Cyclone Separator

Seyed Soheil Mousavi Ajarostaghi [1*], Seyed Sina Mousavi [2], and Chandrasekhar Bhojaraju [3]

[1,*] Department of Mechanical Engineering, Babol Noshirvani University of Technology, Babol 47148-71167, Iran, soheilmousavi67@gmail.com
[2] Department of Civil Engineering, Babol Noshirvani University of Technology, Babol 47148-71167, Iran; seyedsina.m@gmail.com
[3] Department of Civil Engineering, St Joseph Engineering College (SJEC), India; chandrasekhar.b@sjec.ac.in

*Corresponding author: soheilmousavi67@gmail.com

Abstract. In this work, the impact of cone diameters on the flow field and separation efficiency in a cyclone separator is examined numerically employing Reynolds Stress Model for eight various geometries of cyclone separators. The motion of solid particle in the flow field is modelled utilizing the Eulerian-Lagrangian method. Three-dimensional simulation of the air flow with solid particles in the cyclone is carried out by a commercial computational fluid dynamics software, ANSYS Fluent 18.2. The outcomes depict that by augmenting the small cone diameter, the maximum tangential velocity, static pressure, and collection efficiency decrease, however, cut-off diameter rises. Moreover, the axial velocity has steady trend throughout the cyclone. In addition, according to the obtained results, increasing the large cone diameter causes an increase in the tangential velocity and static pressure and improves the collection efficiency. While, the higher large cone diameter, the lower cut-off diameter. Growing the cyclone's cone small diameter leads to a decrease in the collection efficiency. At a constant particle diameter, four microns, as the cone small diameter ratio augments by 150%, collection efficiency declines by about 10.42%. Augmentation the cyclone's cone small diameter causes an increase in the cut-off diameter. By increasing the cone small diameter by about 150%, the cut-off diameter value increases by about 75%. Furthermore, the collection efficiency rises by growing the cyclone's cone large diameter. At a constant particle diameter, four microns, as the cone large diameter augments by 50%, the collection efficiency increases by about 18.75%. Also, in all considered cone large diameter, the collection efficiency increases as the particle diameter rises. The cut-off diameter decreases by growing the cyclone's cone large diameter. By augmenting the cone large diameter by about 50%, the cut-off diameter value declines by about 57.69%.

Keywords: GAS Cyclone Separator, Eulerian–Lagrangian, Reynolds Stress Model, Separation Efficiency, Numerical Simulation.

1. Introduction

Gas cyclone separators (GCSs) are extensively employed in various industries to collect dust from gas streams for a specific goal. They are suitable for industrial dedusting in high pressure and temperature because of their geometrical simplicity, relative economy in consumption, and flexibility [1].

Cyclones are air pollution abatement tools for suspended particles from air flows in various industries. Cyclones have low initial costs, maintenance requirements, and energy feeding. Centrifugal force is utilized by a cyclone to collect particles from the air flow. In a cyclone, a fluid flow full-filled of solid particle as dust, enters tangentially and spirals downward. Then, the centrifugal and inertial forces compel the particles to the cyclone's walls which the particles slide down to the bottom of cone and finally the particles are collected. The airflow moves to the end of the cyclone, then it reverses direction and flows to the out of the device top [2].

The key mechanism for the separation of phases with various densities in a GCS is extremely turbulent flow. The Stairmand is an efficient cyclone with a typical geometric design. The flow field in the GCS is a very complex three-dimensional swirling flow.

Numerous studies [3-7] have been presented to examine the effects of various geometrical factors, including; cyclone length and pipe geometry of inlet and outlet ports. Regarding the performance of cyclones, little works have been done on the geometry of the cone section.

Elsayed and Lacor [8] evaluated numerically the fluid flow and particle collection performance inside a GCS with four various geometries of the dust outlet. Their outcomes demonstrated that there was no significant change in the maximum tangential velocity in various cases. Moreover, it was recorded that radial acceleration was not achieved inside the considered GCS. Azadi et al. [9] presented a 3D Eulerian–Lagrangian CFD model inside a GCS to examine the impacts of the geometrical (cyclone size) and fluid flow (inlet velocity) parameters on the fluid flow. The evaluated output factors were pressure drop and cut-off diameter. The employed turbulence models were Reynolds stress model (RSM) and k–ε RNG. Moreover, the obtained numerical results demonstrated that among the considered turbulence models, RSM is a promising model to model the swirl flows inside the GCS.

Souza et al. [10] studied the flow field inside a small GCS to evaluate the influences of the shape and length of the outlet duct on the pressure drop and grade efficiency. The Large Eddy Simulation (LES) was employed to model the fluid flow inside the GCS and Lagrangian description-based on Newton's second law of the solid particle was utilized to model particle dispersion. They examined more than thirty cases with various outlet duct configurations. Obtained outcomes revealed that the pressure drop first declines by increasing the length of the duct until it reaches a minimum and from this point on, it rises uniformly by growing the length of the duct. In the other work, Kozołub et al. [11] analyzed the fluid flow inside a GCS under high particle loads by performing numerical simulation. The RSM turbulence model was employed for simulation of the complex swirl flows inside the proposed GCS. The considered factors in this work include three inlet velocities and multiple particle loads. Furthermore, the employed model was validated with the experimental results and the comparison analysis showed acceptable agreement between the proposed numerical results and experimental outcomes.

A numerical work based on the traditional Lapple GCS has been provided by Pei et al. [12] in which some blades have been inserted in the vortex tube. The RSM and Lagrangian particle tracking model (LPT) have been employed to model the turbulence gas flow and particle dispersion, respectively. The outcomes depicted that the growth of the vortex core diameter leads to an increase in the pressure lose. It is worth mentioning that the pressure in the vortex finder directly related to the pressure loss in the GCS. Also, utilizing the proposed blades cause about 16.41 percent reduction in the pressure loss and about 0.64 percent augmentation in the separation efficiency.

The main objective of present work is to investigate the effects of cone diameters (small and large) of GCS on the performance factors and gas flow field by performing the numerical analysis, employing CFD commercial package. The considered GCSs with various values of small and large diameters of the cone have been analyzed numerically using ANSYS Fluent 18.2. To simulate the complex fluid flow of swirl gas flows inside the GCS, the RSM, as an efficient turbulence model, has been employed. Moreover, to model the particle dispersion inside the proposed GCS, the discrete phase model (DPM) has been utilized. The impacts of the cone diameters, including; small and large one, on the velocity field, pressure drop, and therefore, cyclone performance have been examined here. As a consequence, two factors have been calculated, including; cut-off diameter and separation efficiency, to analyze the proposed GCS.

2. Mathematical Modeling

2.1. Turbulence Model

Gas-solid fluid flow inside the GCS is extremely swirling and complex and in this kind of specific fluid flow, selection of the suitable and precise turbulence model is challenging. There are various types of turbulence models, like standard k–ε model, renormalization group (RNG) k–ε model, and RSM. Several works presented that RSM is very appropriate in modelling the specific gas flow field inside the GCS [13-15]. The RSM turbulence model has been employed here to investigate the impacts of the GCS cone diameters on the gas-particle fluid flow inside the proposed GCS. In the utilized numerical model (RSM), as well as the continuity and momentum equations (totally four equations), three transport equations for the turbulent normal stresses and three transport equations for the turbulent shear stresses are solved. Furthermore, additional transport equations for the turbulent dissipation rate (ε) and turbulent kinetic energy (k) are considered and calculated [16].

2.2. Configurations of Considered Cyclones

The schematic of the investigated GCS is illustrated in Figure 1. The geometrical dimensions of the tested GCSs are listed in Table 1. The numerical simulations were performed on eight GCSs. Four GCSs (d_0/D =0.2 - 0.5) were considered to present the influence of the cyclone's cone small diameter on the separation efficiency of the GCS. The effect of cyclone's cone large diameter on the separation efficiency was analyzed by considering four GCSs (D_0/D = 0.8-1.2).

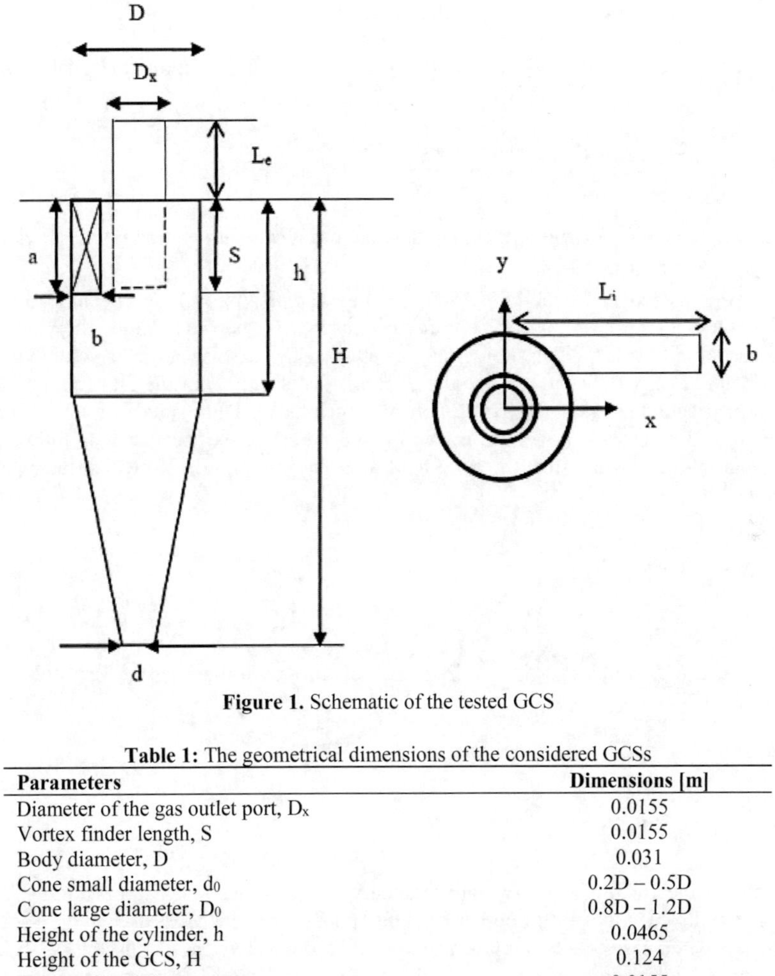

Figure 1. Schematic of the tested GCS

Table 1: The geometrical dimensions of the considered GCSs

Parameters	Dimensions [m]
Diameter of the gas outlet port, D_x	0.0155
Vortex finder length, S	0.0155
Body diameter, D	0.031
Cone small diameter, d_0	0.2D – 0.5D
Cone large diameter, D_0	0.8D – 1.2D
Height of the cylinder, h	0.0465
Height of the GCS, H	0.124
Height of the inlet port, a	0.0155
Width of the inlet port, b	0.0081375

2.3. The Governing Equations

The continuity equation for the incompressible fluid flow is as follows [17-19]:

$$\frac{\partial \overline{u_i}}{\partial x_i} = 0 \quad (1)$$

Balance of momentum is given as follows [17-19]:

$$\frac{\partial \overline{u_i}}{\partial t} + \overline{u_j}\frac{\partial \overline{u_i}}{\partial x_j} = -\frac{1}{\rho}\frac{\partial \bar{P}}{\partial x_i} + \vartheta\frac{\partial^2 \overline{u_i}}{\partial x_j \partial x_i} - \frac{\partial}{\partial x_j} R_{ij} \tag{2}$$

where $R_{ij} = \overline{\acute{u}_i \acute{u}_j}$ is the Reynolds stress tensor. $\acute{u}_i = u_i - \overline{u_i}$ is the ith fluctuating velocity component. The differential transport equation for calculation of the turbulence stress components is as follows [17-19]:

$$\frac{\partial}{\partial t}R_{ij} + \overline{u_k}\frac{\partial}{\partial x_k}R_{ij} = \frac{\partial}{\partial x_k}\left(\frac{\vartheta_t}{\sigma^k}\frac{\partial}{\partial x_k}R_{ij}\right) - \left[R_{ik}\frac{\partial \overline{u_j}}{\partial x_k} + R_{jK}\frac{\partial \overline{u_i}}{\partial x_k}\right] - C_1\frac{\varepsilon}{k}\left[R_{ij} - \frac{2}{3}\delta_{ij}k\right] - C_2\left[P_{ij}\frac{2}{3}\delta_{ij}P\right] - \frac{2}{3}\delta_{ij}\varepsilon \tag{3}$$

$$P_{ij} = -\left[R_{iK}\frac{\partial \overline{u_j}}{\partial x_k} + R_{jK}\frac{\partial \overline{u_i}}{\partial x_k}\right] \tag{4}$$

$$P = P_{ij}\frac{1}{2} \tag{5}$$

where μ_t is the turbulent (eddy) viscosity; and $\sigma^k = 1$, $C_1=1.8$, $C_2=0.6$ are empirical constants. The turbulence dissipation rate (ε) equation is as follows [20]:

$$\frac{\partial \varepsilon}{\partial t} + \overline{u_j}\frac{\partial \varepsilon}{\partial x_j} = \frac{\partial}{\partial x_j}\left[\left(\vartheta + \frac{\vartheta_t}{\sigma^\varepsilon}\right)\frac{\partial \varepsilon}{\partial x_j}\right] - C^{\varepsilon 1}\frac{\varepsilon}{K}R_{ij}\frac{\partial \overline{u_i}}{\partial x_j} - C^{\varepsilon 2}\frac{\varepsilon^2}{K} \tag{6}$$

where $K=\frac{1}{2}\overline{\acute{u}_i \acute{u}_i}$ is the fluctuating kinetic energy, and ε is the turbulence dissipation rate. The values of constants are $\sigma^\varepsilon=1.3$, $C^{\varepsilon 1}=1.44$, and $C^{\varepsilon 1}=1.92$.

In the present work, discrete phase model (DPM), as a Lagrangian approach, has been employed to model the particle dispersion through the gas flow as continuum phase. On the other hand, the employed model for simulation of the gas-solid flow is a Eulerian–Lagrangian approach. The solid particles volume fraction is defined as the ratio of the volume of particles to the volume of fluid. The accuracy of the DPM model is high when the solid particles volume fraction is less than 10 to 12 percent which makes DPM suitable for modeling particles full-filled flows. The amount of particle in a GCS is low (3 to 5 percent), so, it can be carefully expected that the occurrence of the particles does not affect the flow field (one-way coupling) [20] which the equation of particle motion is as follows [2]:

$$\frac{du_p}{dt} = \frac{18\mu}{\rho_p d_p^2}\frac{C_d Re_p}{24}(u_i - u_{pi}) + g_i\frac{(\rho_p - \rho)}{\rho_p} \tag{7}$$

$$\frac{dx_{pi}}{dt} = u_{pi} \tag{8}$$

where index p refers to the particle. C_D is the drag coefficient. Re_p is the relative Reynolds number which is calculated as follows [17]:

$$Re_p = \frac{\rho_p d_p (u - u_p)}{\mu} \tag{9}$$

2.4. Boundary Conditions

At the inlet port of the GCS, a *Velocity_Inlet* boundary condition has been utilized. An *Outflow* boundary condition has been employed at the port of outlet. For other boundaries, the *No_Slip* boundary condition has been employed. In all considered cases, the volume flowrate of the air has been kept constant by 50 L/min. Also, the characteristic length and turbulent intensity are 0.07 times of the inlet width and 5 percent, respectively.

2.5. Numerical Schemes

The Pressure Staggered Option (PRESTO) as pressure interpolation scheme has been employed because of the high swirl flow and high pressure gradient in the considered problem. Also, the Semi–Implicit Method Pressure-Linked Equations Consistent (SIMPLEC) algorithm has been utilized for the pressure-velocity coupling. Moreover, Quadratic Upstream Interpolation for Convective Kinetics (QUICK) scheme has been used for equations of momentum.

2.6. Grid Independence and Validation Evaluations

2.6.1 Grid Independence Analysis

To save the time and cost of the simulation, performing the mesh independence analysis is necessary in any numerical work. The results were obtained for four various grids, including; 153,000, 353,000, 503,000, and 706,000 cells. These results were compared with each other as shown in Figure 2 considering three factors including; static pressure, tangential velocity, and axial velocity. It has been achieved that there is no major variance among the tested grids. It was due to time saving that the grid with 153,000 cells is selected for performing numerical solution.

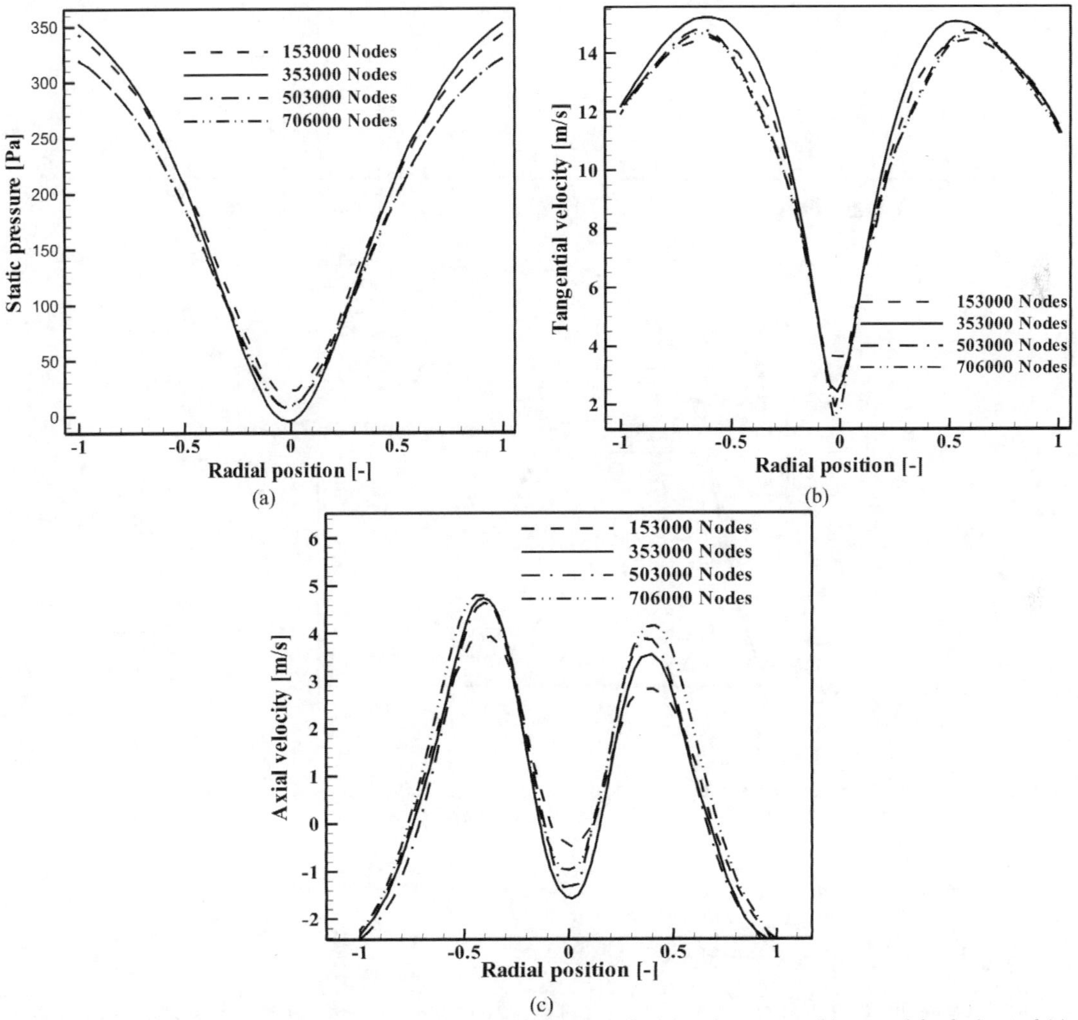

(c)

Figure 2. The comparison between the results of various grids based on the (a) static pressure, (b) tangential velocity, and (c) axial velocity.

2.6.2 Validation Analysis

The results of the investigation of Elsayed and Lacor [21] has been employed to validate the present numerical model. The outcomes of the present work are compared with the outcomes of the reference work considering three parameters, including; static pressure, tangential velocity, and axial velocity curves at a specific axial station as shown in Figure 2 (please check the Elsayed and Lacor [21] for additional details about the employed GCS in the verification analysis). Numerical results show good agreement with the reference work.

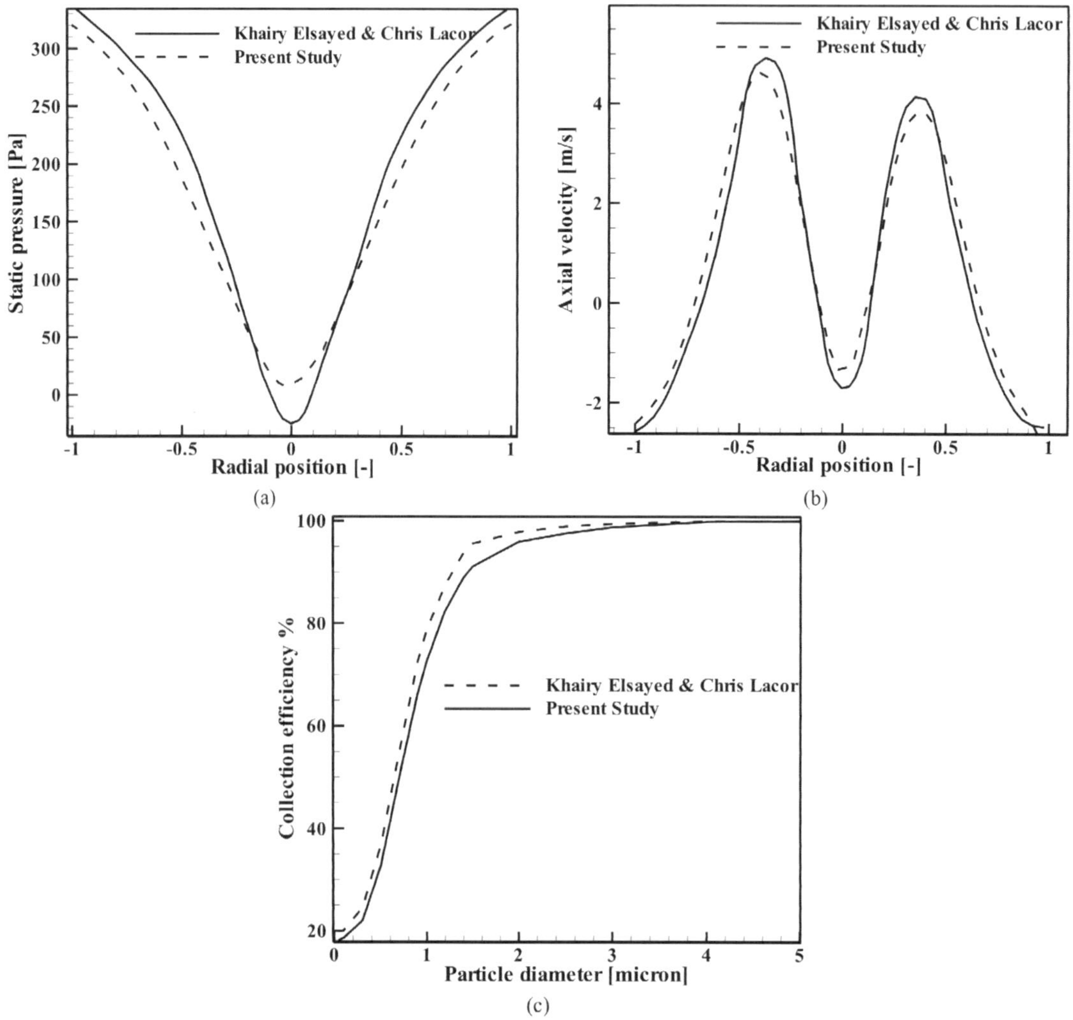

Figure 3. Comparison of the curves between the obtained outcomes of the employed RSM and results of the Elsayed and Lacor [21]; (a) static pressure, (b) axial velocity, and (c) collection efficiency.

3. Results and Discussion

3.1. The Impact of the Small Diameter of the GCS Cone

In order to examine the influence of cyclone's cone small diameter on the performance characteristics such as the collection efficiency, cut-off diameter, static pressure, and axial and tangential velocity curves, the obtained numerical outcomes have been presented in Figures (4-7). Figure 4 demonstrates the contours of the static pressure for GCSs with various cone small diameter. It can be seen that, the non-symmetry of the fluid flow rises by augmentation the cone small diameter. Besides, the maximum value of the static pressure declines by augmentation of the cone small diameter.

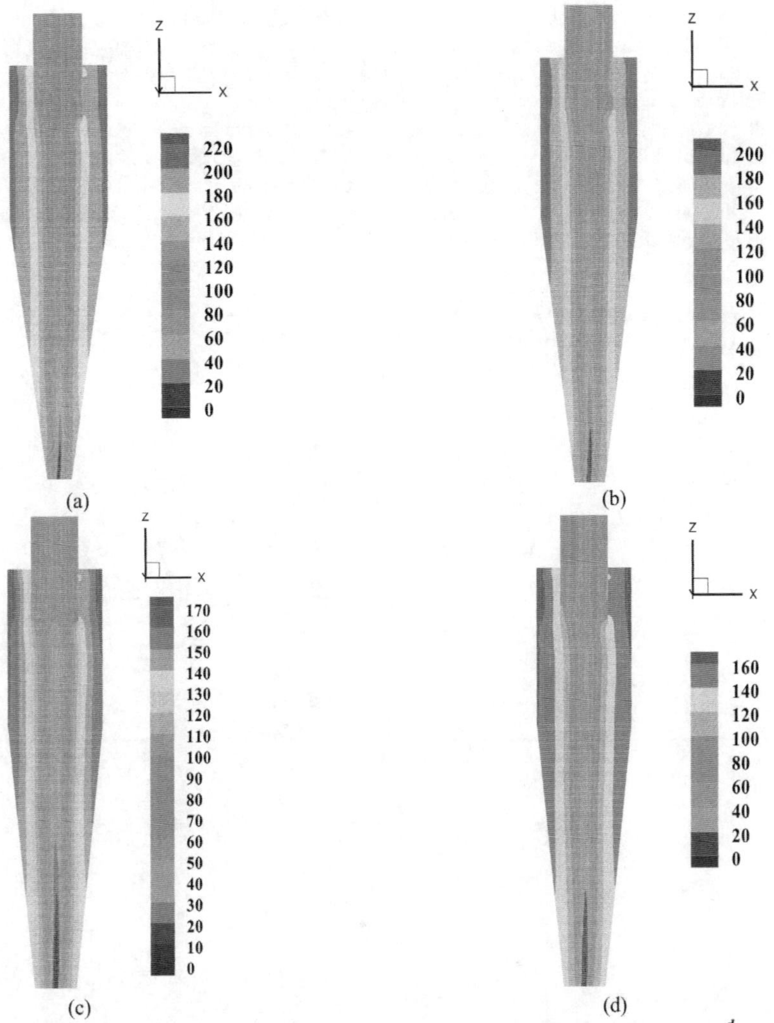

Figure 4. The contours of the time averaged static pressure for various cone small diameter at Y = 0; (a) $\frac{d_0}{D} = 0.2$ (b) $\frac{d_0}{D} = 0.3$ (c) $\frac{d_0}{D} = 0.4$ (d) $\frac{d_0}{D} = 0.5$

The static pressure and axial and tangential velocity curves have calculated at various cone small diameter of the GCS. Figure 5 present the radial curves of the time averaged static pressure, tangential and axial velocity at a specific axial station. The radial curves of statics pressure for GCSs with various cone small diameter are demonstrated in Figure 5a. At a certain axial position, by increasing the cyclone's cone small diameter, the static pressure decreases (Figure 5a). It should be noted that the differences between cases $d_0/D = 0.2$ and $d_0/D = 0.3$ are very low. Also, it can be seen that in all presented curves, the minimum static pressure belongs to the dimensionless radial position of the zero. At this point, for instance, by increasing the cone small diameter ratio from $d_0/D = 0.2$ to $d_0/D = 0.5$ (150% growth), the static pressure declines by about 47.17%.

The axial velocity curves of GCSs with various cone small diameter are shown in Figure 5b. The axial velocity patterns for all tested GCSs illustrate the shape of an inverted "W" and there is no weighty change in axial velocity radial curve for GCSs with various cone small diameters (Figure 5b). The tangential velocity curves for various models with different cone small diameters are demonstrated in Figure 5c. Accordingly, it can be seen that similar to the axial velocity curve (Figure 5b), the tangential velocity curves for all tested models show the shape of an inverted "W". The tangential velocity curves reveal the so-called *Rankine* vortex, which includes two sections, (i) an outer free vortex and (ii) an inner solid rotation in the center (Figure 5c).

Figure 5c depicts that the distribution of the tangential velocity in the inner area is slightly identical at various parts for the similar GCS. In the outer area, because of the severe slump in velocity magnitude near wall area, the distribution is dissimilar and the variation in the highest tangential velocity is relatively restricted. Tangential velocity in the outer area of GCSs declines when the cyclone's cone small diameter is augmented (Figure 5c). It should also be noted that the tangential velocity curves for all models on both sides of the zero dimensionless radial position line have symmetry, which the same results are observed in Figure 5a for static pressure curves however, it is not observed in Figure 5b for axial velocity. For instance, at dimensionless radial positions of -0.5 and 0.5 (the values are the same because of the presence of symmetry in the tangential velocity curves), as the cone small diameter ratio rises from $d_0/D = 0.2$ to $d_0/D = 0.5$ (150% growth), tangential velocity decreases by about 15.12%.

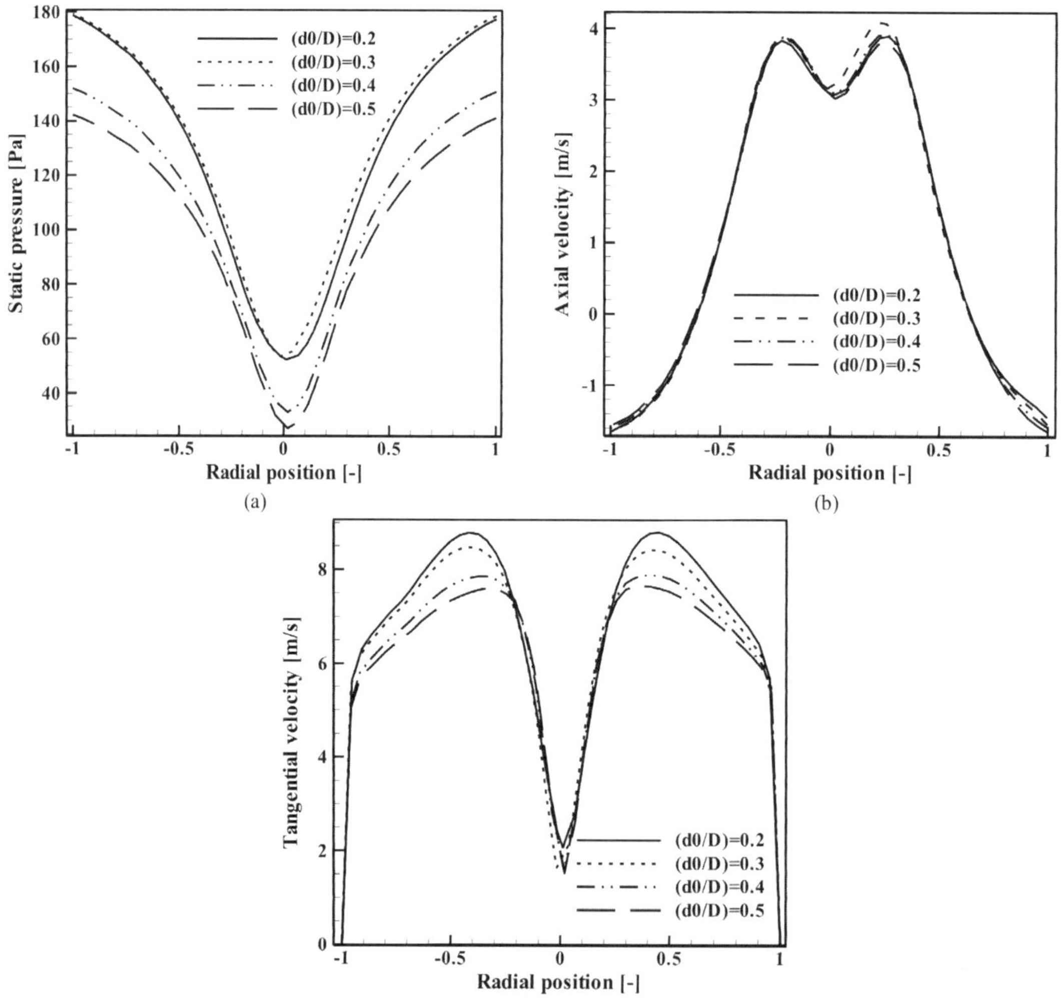

Figure 5. (a) The radial curve for the static pressure at Z = 0.05425 m from the GCS bottom and plane X =0, (b) The radial curve for the axial velocity at Z = 0.05425 m from the GCS bottom and plane X =0, and (c) The radial curve for the tangential velocity at Z = 0.05425 m from the GCS bottom and plane X =0.

Figure 6 demonstrates the contours of the time averaged axial velocity for GCSs with various cone small diameters. Accordingly, by increasing the cone small diameter, non-symmetry of the fluid flow rises. Moreover, higher value of the cone small diameter causes higher value of the axial velocity in the middle area of the GCS, however, there is no significant change of axial velocity in the area near the walls for various cone small diameters.

IOP Conf. Series: Earth and Environmental Science 1149 (2023) 012005 doi:10.1088/1755-1315/1149/1/012005

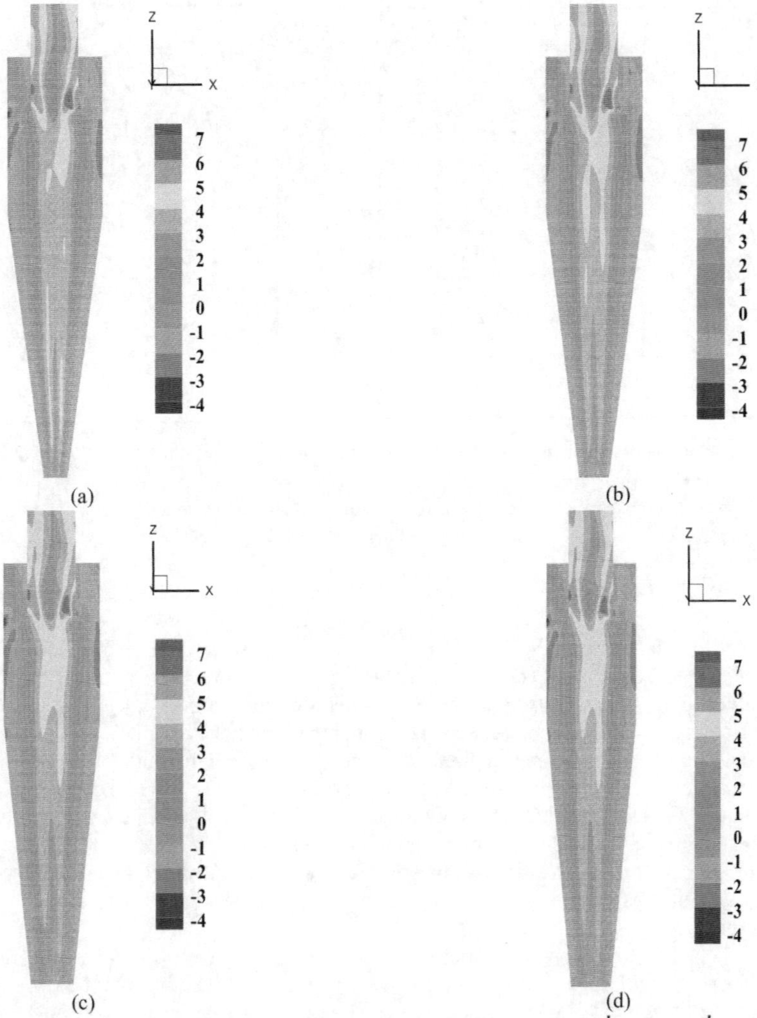

Figure 6. The contours of the axial velocity for various cone small diameter at Y = 0; (a) $\frac{d_0}{D}$ = 0.2 (b) $\frac{d_0}{D}$ = 0.3 (c) $\frac{d_0}{D}$ = 0.4 (d) $\frac{d_0}{D}$ = 0.5

To evaluate the impact of cyclone's cone small diameter on the performance parameters, the collection efficiency and cut-off diameter have been examined and plotted in Figure 7a and Figure 7b, respectively. Accordingly, it can be realized that growing the cyclone's cone small diameter decreases the collection efficiency (Figure 7a) at the cost of augmenting the cut-off diameter (Figure 7b). Tangential velocity in the outer area of GCSs declines when the cyclone's cone small diameter is augmented (Figure 5c) which leads to the lower separation efficiency obtained in GCSs at higher cone small diameter (Figure 7a).

For instance, by considering Figure 7a at a constant particle diameter, 4 microns, as the cone small diameter ratio augments from d_0/D = 0.2 to d_0/D = 0.5 (150% growth), collection efficiency declines by about 10.42%. Furthermore, according to the Figure 7b, by augmentation of the cone small diameter ratio by about 150% (from d_0/D = 0.2 to d_0/D = 0.5), the cut-off diameter value increases by about 75%.

Moreover, Figure 7a demonstrates that in all considered cone small diameter, the collection efficiency increases as the particle diameter rises. For instance, at d_0/D = 0.5, as the particle diameter augments from 2 to 6 microns (200% growth), the collection efficiency rises by about 50%.

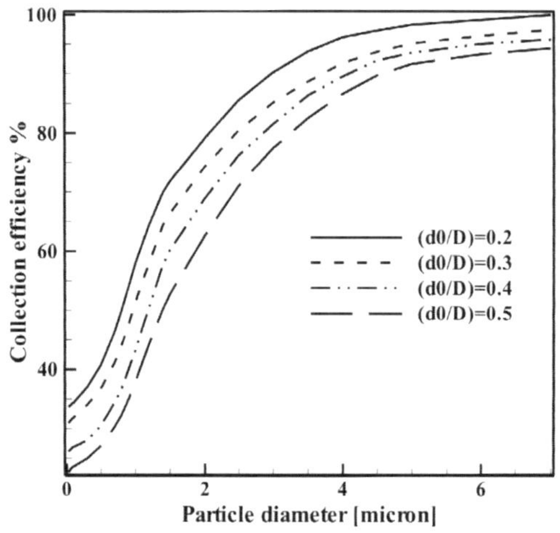

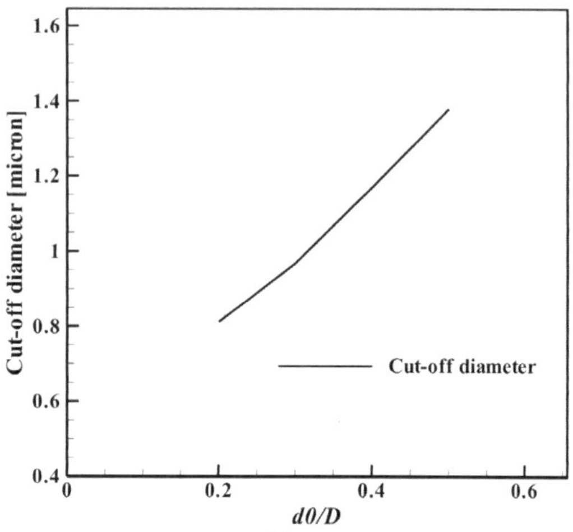

Figure 7. (a) The collection efficiency curve, and (b) the cut-off diameter curve for various cone's small diameter and particle diameter.

3.2. The Impact of the Large Diameter of the GCS Cone

To evaluate the influence of cyclone's cone large diameter on the separation performance of the GCS, the static pressure, tangential, and axial velocity curves for GCSs with various cone large diameters are compared in Figure 8. The radial curves of the statics pressure for GCSs with various cone large diameters are demonstrated in Figure 8a. Accordingly, at a certain dimensionless radial position, by growing the cyclone's cone large diameter, the static pressure increases. For instance, at dimensionless radial position of zero, the static pressure of the case $D_0/D = 1.2$ is 11.67 times more than the static pressure of the case $D_0/D = 0.8$. It is worth mentioning that the differences between the obtained results of cases $D_0/D = 1.1$ and $D_0/D = 1.2$ are not significant. Also, Figure 8a shows that the curves of the static pressure versus the dimensionless radial position are "V" shape which means that the minimum value of static pressure belongs to the dimensionless radial position of zero for all considered cone large diameter. Moreover, Figure 8a depicts that the static pressure curves are symmetric with respect to the zero dimensionless radial position line.

Figure 8b depicts the radial curves of axial velocity for GCSs with various cone large diameters. The axial velocity curve for GCSs with higher cone large diameter, D_0= (0.9-1.2)D, illustrate the shape of an inverted "W" shape while that of GCS with D_0 = 0.8D illustrates an inverted "V" shape curve (Figure 8b). For all models except case D_0 = 0.8D, the minimum axial velocity belongs to the dimensionless radial position of zero. Furthermore, Figure 8b demonstrates that by increasing the cone large diameter, the axial velocity magnitude in the inner area of the cyclone (dimensionless radial position between about -0.5 and 0.5) declines while the axial velocity magnitude in the outer area of the cyclone (dimensionless radial position below about -0.5 and more than about 0.5) rises. For instance, at the zero dimensionless radial position, 50% growth in the cone large diameter (from D_0 = 0.8D to D_0 = 1.2D) leads to about 68% decline in the axial velocity magnitude. However, at dimensionless radial position equal to -1, by increasing the cone large diameter from D_0 = 0.8D to D_0 = 1.2D, 50% growth, the axial velocity magnitude augments by about 52.17%. It is important to note that, as seen in the Figure 8b, the axial velocity curve is not symmetrical, unlike the static pressure curve (Figure 8a).

The tangential velocity curves for GCSs with various cone large diameters are illustrated in Figure 5c. The tangential velocity distribution fluctuates only roughly with axial locations for the similar GCS. In other words, if the tangential velocity rises at one part of the GCS for one case compared to the others, it increases at all other parts. The tangential velocity curve is very similar for all GCSs (Figure 5c). The minor differences between the various cases can be distinguished only in the area of the dimensionless radial position between about -0.5 and 0.5. In this area, the maximum and minimum tangential velocity magnitude belong to the cases D_0 = 1.2D and D_0 = 0.8D, respectively. The tangential velocity curves are not symmetrical similar to the axial velocity curves.

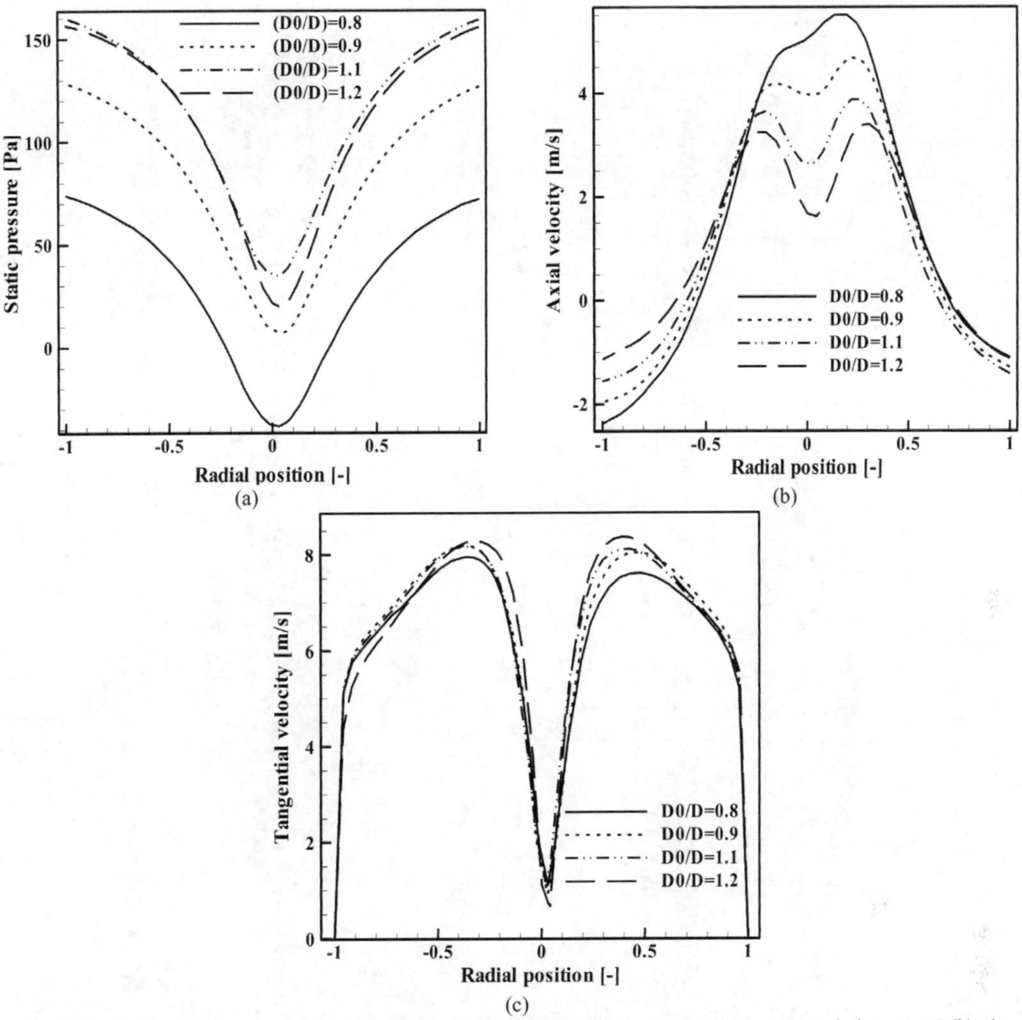

Figure 8. (a) The radial curve for the static pressure at Z = 0.05425 m from the GCS bottom and plane X = 0 (b), the radial curve for the axial velocity at Z = 0.05425 m from the GCS bottom and plane X = 0, and (c) the radial curve for the axial velocity at Z = 0.05425 m from the GCS bottom.

Figure 9 demonstrates the contours of the time averaged static pressure for tested GCSs. Accordingly, it can be realized that by increasing in the cone large diameter, the static pressure rises and the lowest static pressure is in the center of the cyclone and the highest static pressure is in the areas near the wall. It can also be seen that with the increase in cone large diameter, more areas around the wall reach the maximum static pressure and more areas in the center reach the minimum static pressure.

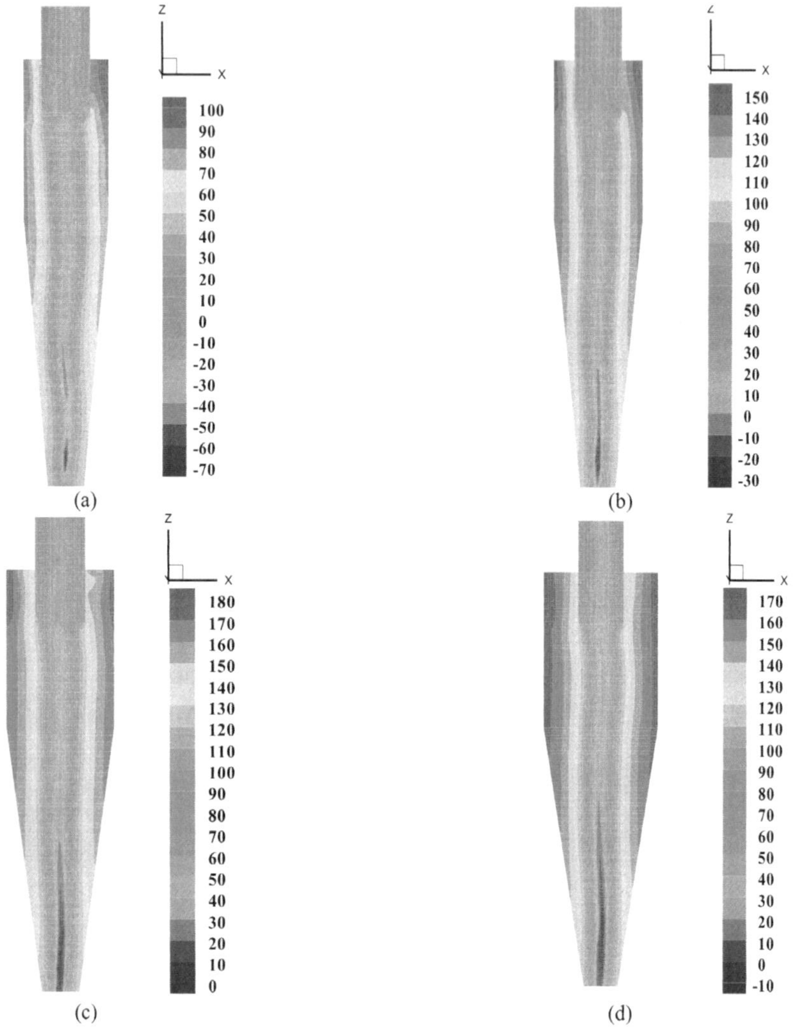

Figure 9. The contours of the time averaged static pressure for various large diameter of separator cone at Y = 0; (a) $\frac{D_0}{D}$ = 0.8, (b) $\frac{D_0}{D}$ = 0.9, (c) $\frac{D_0}{D}$ = 1.1, and (d) $\frac{D_0}{D}$ = 1.2.

In order to investigate the influence of GCS cone large diameter on the separation efficiency of GCS, the collection efficiency and cut-off diameter curves are shown in Figure 10a and Figure 10b, respectively. As is clear from Figure 10a, the collection efficiency rises by growing the cyclone's cone large diameter, consequently, the cut-off diameter decreases by growing the cyclone's cone large diameter (Figure 10b).

For instance, by considering Figure 10a at a constant particle diameter, 4 microns, as the cone large diameter ratio augments from $D_0/D = 0.8$ to $D_0/D = 1.2$ (50% growth), collection efficiency increases by about 18.75%. Furthermore, according to the Figure 10b, by augmentation of the cone large diameter ratio by about 50% (from $D_0/D = 0.8$ to $D_0/D = 1.2$), the cut-off diameter value declines by about 57.69%.

Moreover, Figure 10a demonstrates that in all considered cone large diameter, the collection efficiency increases as the particle diameter rises, similar to the curves of Figure 7a for cone small diameter. For instance, at $D_0/D = 0.8$, as the particle diameter augments from 2 to 6 microns (200% growth), the collection efficiency rises by about 157.14%.

IOP Conf. Series: Earth and Environmental Science 1149 (2023) 012005 doi:10.1088/1755-1315/1149/1/012005

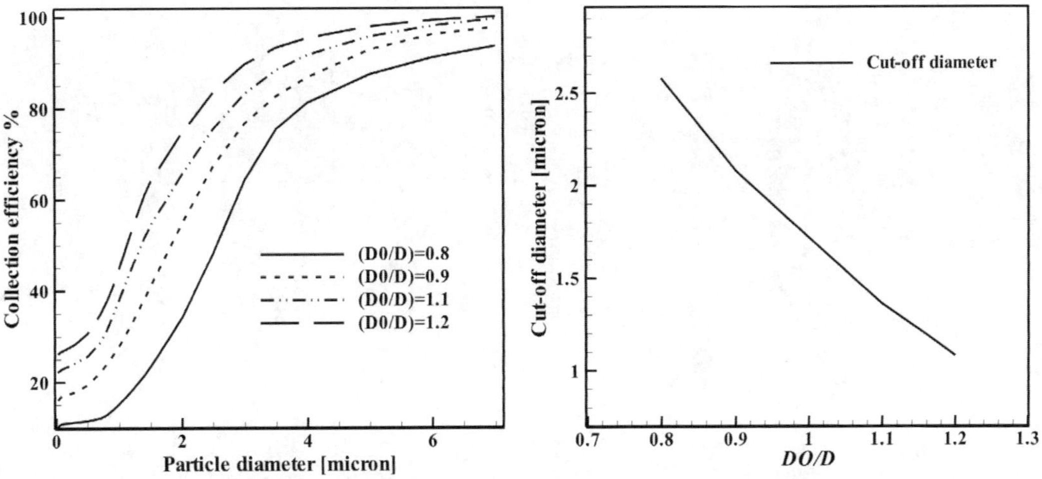

(a) (b)

Figure 10. (a) The separation efficiency curves, and (b) the cut-off diameter curves for various cone large diameters and particle diameters.

Axial velocity contours at a middle slice inside the computational area (Y = 0) are shown in Figure 11. As can be seen, in the central area of the GCS, an augmentation in the cone large diameter of the separator cone causes a decline in the axial velocity. Also, results show that in the area near the walls, growing the large diameter of the separator cone causes an increase in the axial velocity.

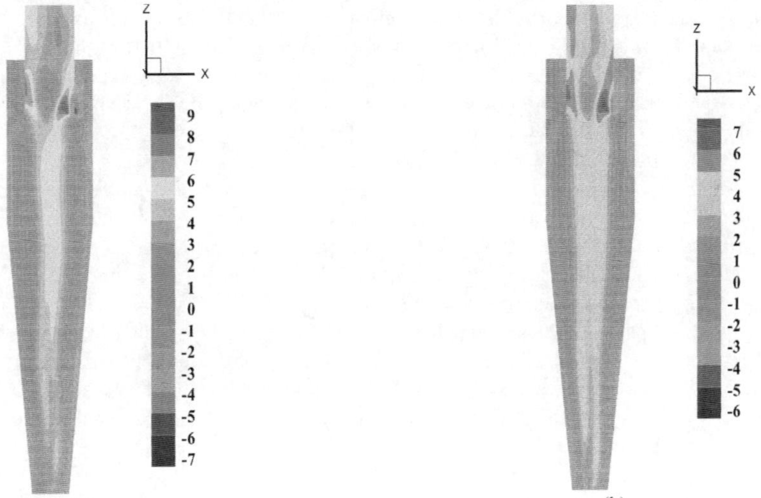

(a) (b)

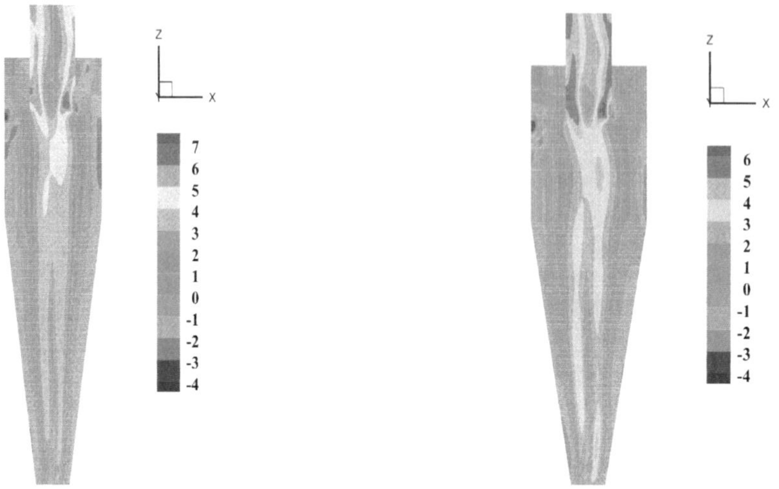

(c) (d)

Figure 11. The contours of the axial velocity for various cone large diameters of separator at Y = 0; (a) $\frac{D_0}{D} = 0.8$ (b) $\frac{D_0}{D} = 0.9$ (c) $\frac{D_0}{D} = 1.1$ (d) $\frac{D_0}{D} = 1.2$

4. Concluding Remarks

A numerical investigation is performed to evaluate the fluid flow and separation performance inside a cyclone separator. The Reynolds stress model (RSM) is employed to model the turbulent flow of air in the cyclone and discrete phase model (DPM) is utilized to track individual particles through the continuum fluid (gas or air). The effects of the cyclone's cone diameters, small and large, on the cyclone separation performance and flow field are numerically evaluated employing a commercial finite volume method code, ANSYS FLUENT 18.2. Based on the presented results, the following conclusions have been obtained:

- Variations of the cyclone's cone diameters affects the shape of velocity curves (axial and tangential) inside the considered cyclone significantly.
- By increasing the cyclone's cone small diameter, the static pressure decreases. The minimum static pressure belongs to the zero dimensionless radial position. At this point, by increasing the cone small diameter by 150%, the static pressure declines by about 47.17%.
- The axial velocity patterns for all tested GCSs illustrate the shape of an inverted "W" and there is no significant change in axial velocity radial curve for GCSs with various cone small diameters.
- Tangential velocity in the outer area of GCSs declines when the cyclone's cone small diameter is augmented. The tangential velocity curves for all models on both sides of the zero dimensionless radial position line have symmetry, which the same results are observed for static pressure curves however, it is not observed for axial velocity.
- At dimensionless radial positions of -0.5 and 0.5, as the cone small diameter rises by 150%, tangential velocity decreases by about 15.12%.
- Growing the cyclone's cone small diameter leads to a decrease in the collection efficiency. At a constant particle diameter, four microns, as the cone small diameter ratio augments by 150%, collection efficiency declines by about 10.42%.
- Augmentation the cyclone's cone small diameter causes an increase in the cut-off diameter. By increasing the cone small diameter by about 150%, the cut-off diameter value increases by about 75%.
- By growing the cyclone's cone large diameter, the static pressure increases. At dimensionless radial position of zero, by increasing the cone large diameter by about 50%, the achieved static pressure is 11.67 times more than the reference case.
- By augmenting the cone large diameter, the axial velocity magnitude in the inner area of the cyclone drops but the axial velocity magnitude in the outer area of the cyclone increases. At the zero dimensionless radial position, 50% growth in the cone large diameter causes about 68% decline in the axial velocity magnitude.

However, at dimensionless radial position equal to -1, by increasing the cone large diameter by 50%, the axial velocity magnitude augments by about 52.17%.

- The collection efficiency rises by growing the cyclone's cone large diameter. At a constant particle diameter, four microns, as the cone large diameter augments by 50%, the collection efficiency increases by about 18.75%. Also, in all considered cone large diameter, the collection efficiency increases as the particle diameter rises.
- The cut-off diameter decreases by growing the cyclone's cone large diameter. By augmenting the cone large diameter by about 50%, the cut-off diameter value declines by about 57.69%.

References

[1]. Hoekstra, A. J., Derksen, J. J., & Van Den Akker, H. E. A. (1999). An experimental and numerical study of turbulent swirling flow in gas cyclones. *Chemical Engineering Science*, *54*(13-14), 2055-2065.
[2]. Cooper and Alley, 2002
[3]. Le, D. K., & Yoon, J. Y. (2020). Numerical investigation on the performance and flow pattern of two novel innovative designs of four-inlet cyclone separator. *Chemical Engineering and Processing-Process Intensification*, 150, 107867.
[4]. Zhang, P., Duan, J., Chen, G., & Wang, W. (2019). Numerical investigation on gas-solid flow in a circumfluent cyclone separator. *Aerosol and Air Quality Research*, 19(5), 971-980.
[5]. Wei, Q., Sun, G., & Gao, C. (2020). Numerical analysis of axial gas flow in cyclone separators with different vortex finder diameters and inlet dimensions. *Powder technology*, 369, 321-333.
[6]. Venkatesh, S., Sivapirakasam, S. P., Sakthivel, M., Ganeshkumar, S., Prabhu, M. M., & Naveenkumar, M. (2021). Experimental and numerical investigation in the series arrangement square cyclone separator. *Powder Technology*, 383, 93-103.
[7]. Misiulia, D., Antonyuk, S., Andersson, A. G., & Lundström, T. S. (2020). High-efficiency industrial cyclone separator: A CFD study. *Powder Technology*, 364, 943-953.
[8]. Khairy Elsayed, Chris Lacor "The effect of the dust outlet geometry on the performance and hydrodynamics of gas cyclones", *Computers & Fluids* 68 (2012) 134–147.
[9]. Azadi, Mehdi, Mohsen Azadi, and Ali Mohebbi. "A CFD study of the effect of cyclone size on its performance parameters." *Journal of hazardous materials* 182, no. 1 (2010): 835-841.
[10]. de Souza, Francisco José, Ricardo de Vasconcelos Salvo, and Diego de Moro Martins. "Effects of the gas outlet duct length and shape on the performance of cyclone separators." *Separation and Purification Technology* 142 (2015): 90-100.
[11]. Kozołub, Paweł, Adam Klimanek, Ryszard A. Białecki, and Wojciech P. Adamczyk. "Numerical simulation of a dense solid particle flow inside a cyclone separator using the hybrid Euler–Lagrange approach." *Particuology* 31 (2017): 170-180.
[12]. Pei, Binbin, Liu Yang, Kejun Dong, Yunchao Jiang, Xusheng Du, and Bo Wang. "The effect of cross-shaped vortex finder on the performance of cyclone separator." Powder Technology 313 (2017): 135-144.
[13]. Park, D., Cha, J., Kim, M., & Go, J. S. (2020). Multi-objective optimization and comparison of surrogate models for separation performances of cyclone separator based on CFD, RSM, GMDH-neural network, back propagation-ANN and genetic algorithm. *Engineering Applications of Computational Fluid Mechanics*, 14(1), 180-201.
[14]. Qiang, L., Jianjun, W., Weiwei, X., & Meng, Z. (2020). Investigation on separation performance and structural optimization of a two-stage series cyclone using CPFD and RSM. *Advanced Powder Technology*, 31(9), 3706-3714.
[15]. Li, J., Wang, T., Zhang, L., Chang, J., Song, Z., & Ma, C. (2020). Multi-objective optimization of axial-flow-type gas-particle cyclone separator using response surface methodology and computational fluid dynamics. *Atmospheric Pollution Research*, 11(9), 1487-1499.
[16]. ANSYS Inc. (2020). *ANSYS Fluent User Guide*, Release 2020R2.
[17]. H. Safikhani, M. Akhavan-Behabadi, M. Shams, M.H. Rahimyan, Numerical simulation of flow field in three types of standard cyclone separators, Advanced Powder Technology 21 (4) (2010) 435–442.
[18]. Norouztabar, R., Mousavi Ajarostaghi, S. S., Mousavi, S. S., Nejat, P., Rahimian Koloor, S. S., & Eldessouki, M. (2022). On the Performance of a Modified Triple Stack Blade Savonius Wind Turbine as a Function of Geometrical Parameters. *Sustainability*, 14(16), 9816.

[19]. Mousavi Ajarostaghi, S. S., Aghanezhad, M., Davudi, H., & Mohammadzadeh Amiri, M. (2021). Numerical evaluation of the heat transfer enhancement in a tube with a curved conical turbulator insert. *International Journal of Ambient Energy*, 1-14.

[20]. B.E. Launder, G.J. Reece, W. Rodi, Progress in the development of a reynolds-stress turbulence closure, Journal of Fluid Mechanics 68 (03) (1975) 537–566.

[21]. K. Elsayed, C. Lacor, Optimization of the cyclone separator geometry for minimum pressure drop using mathematical models CFD simulations, Chemical Engineering Science 65 (22) (2010) 6048–6058.

Numerical Analysis of Double Stack Blade Savonius Wind Turbine with Secondary Blades

Seyed Soheil Mousavi Ajarostaghi [1*], Seyed Sina Mousavi [2], and Chandrasekhar Bhojaraju [3]

[1,*] Department of Mechanical Engineering, Babol Noshirvani University of Technology, Babol 47148-71167, Iran, soheilmousavi67@gmail.com
[2] Department of Civil Engineering, Babol Noshirvani University of Technology, Babol 47148-71167, Iran; seyedsina.m@gmail.com
[3] Department of Civil Engineering, St Joseph Engineering College (SJEC), India; chandrasekhar.b@sjec.ac.in

*Corresponding author: soheilmousavi67@gmail.com

Abstract. In this work, the performance of the Savonius vertical axis wind turbine with secondary blades was investigated numerically. The impacts of two geometric parameters of the secondary blades, including the height of secondary blades (h) and the distance between the main and the secondary blades (b) on the output torque were analyzed. Numerical simulations were performed by a commercial CFD code, ANSYS FLUENT 18.2. Numerical results show that employing secondary blades has a constructive impact on the output torque compared to the case without secondary blades. Moreover, the studied geometric parameters of the secondary blades have a significant impoact on the performance of the proposed turbine. It can be seen that by increasing the height of secondary blades, the angular position of the maximum torque is shifted. At b = 2.5 mm, the difference between the models is more comparable. Models h = 6 and 12 mm have the highest output torque, although with a slight difference between them. Model h = 12 mm has the highest output torque. At b = 7.5 mm, the difference between the models is more comparable than b = 2.5 mm. Model h = 12 mm has the highest output torque and models h = 3 and 6 mm are in the next levels. Obtained results in the second section demonstrate that by growing the distance between the secondary and the main blades, in the case of h = 3 mm, the produced torque declines. Furthermore, by increasing the distance between the secondary and the main blades from b = 0 mm to b = 2.5 mm and b = 7.5 mm, the amount of torque produced decreases by 17.33 and 26.66%, respectively. Moreover, by augmenting the distance between the secondary and main blades from b = 2.5 mm to b = 7.5 mm (200% growth), the produced torque decreases by 11.3%.

Keywords: Wind Energy, Vertical Axis Wind Turbine (VAWT), Savonius, Computational Fluid Dynamics (CFD), Double Stacks Blade, Torque

1. Introduction

Today, fossil energies are widely used in various industries, but due to the lack of these energies and the environmental pollution caused by their use, societies have turned to using clean and renewable energies such as solar energy [1-3], wind energy [4], and geothermal energy [5-7]. The use of these energies requires comprehensive and complete information regarding the design and operation of renewable energy converters. Wind energy, like other renewable energy sources, is widely available but scattered. The uneven radiation of the sun at different latitudes on the uneven surface of the earth causes changes in temperature and pressure, and as a result, the wind is created. In addition, the earth's atmosphere, due to its rotation, transfers heat from tropical regions to polar regions, which causes wind.

Wind energy has a fluctuating and intermittent nature and does not blow constantly. Wind energy is considered one of the clean energy technologies because it leaves only a part of nature and the environment. Wind power plants do not produce any type of air pollution or greenhouse gas. The use of wind energy in wind turbines that are used to generate electricity is of the type of fast horizontal axis turbines. The cost of building a wind turbine with a certain diameter increases if the number of blades increases. In places where the power supply network is weak and there are local winds near wind generators, the use of this energy carrier will be more useful. During the industrial revolution, fossil fuels replaced wind energy due to their cheapness and high reliability. Nevertheless, the oil crisis created new tendencies in the field of wind energy technology to generate electricity connected to the grid, pump water and supply electric energy to remote areas. In recent years, environmental problems and the issue of global climate change due to the use of conventional energy sources have intensified these interests [8].

A wind turbine is a turbine that is used to convert the kinetic energy of the wind into mechanical energy, which is called wind power. Wind turbines are made in two types horizontal axis and vertical axis. Small wind turbines are used for applications such as charging batteries or auxiliary power in yachts, while larger wind turbines are used as a source of electrical energy by turning a generator. Other types of wind turbines are used to pump water, which are called wind pumps, or used to grind wheat, which are called windmills, and others [9].

Although there are various designs for wind turbines, they are mainly divided into two general categories based on the direction of the rotation axis: Horizontal Axis Wind Turbines (HAWTS) and Vertical Axis Wind Turbines (VAWTS) [10].

HAWTS have complex construction and their installation is only economically valuable in areas with constant wind and high velocity. But vertical axis turbines have a very simple construction and installation and can be used in different wind velocitys and directions. Unlike horizontal axis turbines, these turbines have low rotation velocity and high torque and are independent of the wind direction. For this reason, they are mostly used to produce electricity for power transmission, including compressed air and hydraulic systems, rather than direct electricity production. These types of turbines are the most common wind turbines that are used. All its components (blades, shaft, generator) are placed on top of a tall tower and the blades are designed to face the wind. The axis of the shaft is horizontal to the ground, and the wind hitting the blades causes them to rotate. At the end of the axis, there is a gear that is connected to the generator and delivers the generated electricity to the power grid. These turbines have several key components that increase efficiency. The vane deflector motor makes the vanes always face the wind. Against strong winds, the turbine has a brake that gives rest to the system and prevents any damage [11].

VAWTS are divided into two main types: (1) Darius wind turbines and (2) Savonius wind turbines [12]. Savonius turbines work like a water wheel with drag while Darius turbines use blades similar to horizontal axis turbines. Vertical axis turbines are placed very close to the ground, which has the advantage of placing its heavy equipment, such as the gearbox and generator, close to the ground, although the wind intensity is lower on the ground, and as a result, less electricity will be produced. Among the other advantages of this type of turbine, we can point out that there is no need for a yaw system, because these types of turbines control the wind from all directions, and this advantage has many advantages over its shortcomings. Among its shortcomings, we can point out that these types of turbines do not start working automatically like horizontal axis turbines. Unlike Darius, the Savonius wind turbine rotates at a low rotational velocity but produces more torque. These turbines with at least two blades connected to a rotating shaft operate with low efficiency but higher reliability than Darius. Most of the usual wind turbines get the power necessary for rotation from the lift force resulting from the aerodynamic shape of the blade. it is wind But Savonius turbines are powered by drag [13,14]. In other words, due to the high and greater drag force in the concave area of the turbine rotors compared to their convex part, the turbine rotates. For this reason, in Savonius turbines, its rotation velocity does not exceed the wind velocity [15].

The use of Savonius turbines, especially in recent decades, has been greatly expanded and many studies have been done to increase their efficiency and optimize their blades. This turbine was developed in 1931 by the Finnish engineer Savonius as an invention for Receipt registration [16]. The advantages of this turbine are easy construction with the lowest cost, low dependence of their operation on the direction of the wind, low noise during operation, low wear of moving parts, diversity in rotor type construction, and high dynamic and static torque [17-19]. Many sources also studied the functional parameters of Savonius wind turbines [20-22]. Also, a lot of research has been done on increasing the efficiency and optimizing the blades of this type of turbine. Altan and Atilgan [23] and Mohamed et al. [24] also conducted experiments in this direction and achieved successful results. Altan and Atilgan [23], in laboratory research along with the numerical analysis of the aerodynamics of the Savonius rotor, succeeded in designing a kind of overlapping rotor plates that led to a 27% increase in the rotor efficiency.

While Mohamed et al. [24] for a lower blade tip velocity range They observed a more than 40% increase in efficiency. The experimental observations made by Kamoji et al. [25] investigated the effect of blade shape in increasing the static torque. Based on a wind tunnel test, Hayashi et al. [26] concluded that wind velocity is a very effective parameter in their design because their design faces problems at high velocitys.

Saha et al. [27] conducted wind tunnel tests for one, two, and three-story Savonius and examined the helical and semi-cylindrical vanes for each. In addition to laboratory research, including the works related to the numerical calculations of Savonius aerodynamics, the following works can be mentioned. Fujisawa et al. [28] investigated pressure contours numerically and compared the results of their work with laboratory work. Kawamura et al. [29] simulated the dynamic torque of a Savonius rotor with the RANS model. Gupta et al. [30] performed a two-dimensional analysis by Fluent software for the aerodynamic analysis of a two-blade Savonius rotor using the K-epsilon method. In the current research, both the experimental study of the performance of the Savonius wind turbine in the wind tunnel and the numerical and three-dimensional analysis, which is the most complete and accurate type of simulation, will be carried out with different numerical methods to achieve a comprehensive analysis of the said wind turbine. In 2013, Hadi Ali [31] conducted a laboratory study of two-bladed and three-bladed Savonius wind turbines. The obtained results showed that the power and torque coefficient in the three-blade mode is better than in the two-blade. In 2013, In 2012, Deb and Gupta [32] investigated and observed a two-bladed wind turbine. The highest static pressure occurs at angles of 90 and 270 degrees. Yoshida and Kawamura [33] investigated the effect of rotor blade curvature on the power factor. The results of this study showed that the greater the depth of the blade, the better the power factor. In 2011, McTavish et al. [34] presented an advanced model of the Savonius wind turbine. In this model, the blade walls are designed in such a way that the rotor produces an additional torque with the contraction of the flow. In 2019, Antar and Elkhoury [35] analyzed the parameters of a Saunius turbine designed with a vertical axis using CFD software in two dimensions. The obtained results showed that in all velocity ratios, there is no general optimal shape for the turbine that has the best performance. And it is different from each other in different situations. In 2020, Ahmed et al. [36] investigated the vertical axis Savonius wind turbine with twisting blades to increase its efficiency. Therefore, the effects of several design parameters such as twist angle, overlap ratio, and blade size ratio along with wind velocity on the performance of Savonius wind turbine have been investigated. They studied the 3D model of the turbine numerically with the help of Fluent software. The results showed that the Savonius rotor with a twist angle of 45 degrees, an overlap ratio of zero, and a blade size ratio of 1.1 obtains the highest net output power compared to other models.

Generally, the purpose of conducting research is to improve performance or design a new idea in the Savonius type of VAWTS. One of the important topics that are always discussed in wind turbines is increasing the positive torque and improving its performance. One of the disadvantages of the Savonius rotor is the creation of a negative torque in the counter-rotating blade. In this research, the goal of increasing the positive torque has been created to overcome the negative torque, and, as a result of increasing the efficiency of the wind turbine, it is Savonius. The main blade of the Savonius turbine is a semi-cylinder, and to improve the performance of this turbine, a secondary blade has been added to it. The geometrical parameters of the proposed wind turbine plays an important role in the amount of the produces torque. So, a comprehensive investigation is required to evaluate the impacts of these parameters on the performance of the proposed VAWTS. The considered geometrical parameters here are (I) the height of secondary blades (h) and (II) the distance between the main and the secondary blades (b).

2. Numerical Study Details

2.1. Definition of the Problem

In the current research, the aim is to numerically simulate the wind turbine rotor for domestic and urban use. Using ANSYS Design Modeler and ANSYS Meshing software, the desired geometries have been created and meshed. As it is known, the conditions of wind flow in the urban environment are different from the mainly windy areas. In windy areas, the wind velocity is high and less turbulent, but in urban areas, the wind flow is turbulent and has a low velocity. Therefore, for extracting energy for small applications, according to the points stated in the *Introduction* section, the Savonius rotor has been selected. One of the features of this rotor is its insensitivity to flow direction and disturbance. This rotor consists of half-cylinders that rotate with the concave part in front of the fluid flow and create the torque required for the rotation of the generator. The problem with this rotor is that when the convex part is placed against the wind flow, it creates a torque opposite to the rotation of the rotor, which greatly reduces the amount of torque produced at certain angles. Therefore, to overcome the created opposing torque, it is possible to increase the favorable torque in some angles to the extent that it overcomes the created opposing torque and transfers a uniform cycle to the generator. For this purpose, a two-stack Savonius rotor was investigated. A secondary blade has been added to the main blade to increase the positive torque. The influences of the height of the second blade (h) and the distance between the center of the main blade and the second blade (b) have been investigated here. The schematic of the computational domain along with the geometrical parameters is shown in Figure 1 and the values of the considered geometrical parameters are given in Table 1. In Table 2, the conditions of rotor geometry and fluid flow are given.

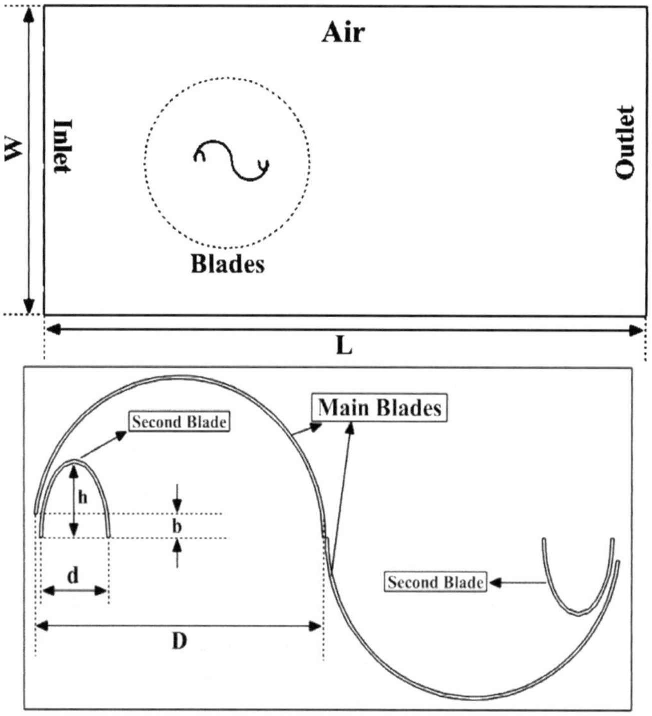

Figure 1. Schematic of the computational domain with geometric parameters and boundary conditions

Table 1: The values of the geometrical parameters of the investigated model

Parameters	Values
The width of the computing area, (W)	0.8 m
The length of the computational area, (L)	1.6 m
Main blade diameter, (D)	0.1 m
The diameter of the second blade, (d)	0.012 m
The distance of the second blade from the main blade, (b)	0 - 7.5 m
The height of the second blade, (h)	3, 6, and 12 m

Table 2: Geometry and flow conditions

Parameter		Value
Flow density [kg/m^3]	ρ	1.22
Inlet velocity [m/s]	V_{inlet}	3-6
Rotor diameter [m]	D	0.2
Number of blades	N	2 and 3
Fluid rotational velocity [Rad/s]	ω	18-40

2.2. Boundary Conditions

In this problem, there are four boundary conditions which are as follow:

- ***Inlet_Velocity***: If the input velocity is known, this condition can be applied to the problem.

IOP Conf. Series: Earth and Environmental Science 1149 (2023) 012006 doi:10.1088/1755-1315/1149/1/012006

- ***Pressure_Outlet***: The outlet pressure is known and equal to the atmospheric pressure.
- ***Inner wall boundary condition***: This condition separates different areas of the computational domain. Despite this condition, the airflow can pass through the common surface between different areas.
- ***Wall boundary condition***: The wall boundary condition is used for the turbine rotor.

2.3. Governing Equations

Continuity:

For the steady-state and compressible conditions, the continuity equation is as follows [37]:

$$\frac{\partial}{\partial x_i}(\rho \bar{u}_i) = 0 \tag{1}$$

Momentum:

For the 2D compressible flow, the Navier-Stokes equation is as follows [37]:

$$\frac{\partial \rho}{\partial_{xi}}(\rho \bar{u}_i \bar{u}_j) = -\frac{\partial \bar{P}}{\partial xi} + \frac{\partial}{\partial xi}[\mu \frac{\partial \bar{u}_i}{\partial x_j} - \rho \bar{u}_i \bar{u}_j] = 0 \tag{2}$$

where $-\rho\, \bar{u}_i \bar{u}_j$ is Reynolds stress. Reynolds stress is modelled for momentum equations as follows [37]:

$$-\rho \bar{u}_i \bar{u}_j = \mu_t [\frac{\partial \bar{u}_i}{\partial x_j} + \frac{\partial \bar{u}_j}{\partial x_i}] - \frac{2}{3} \delta_{ij} [\rho k + \mu_t \frac{\partial \bar{u}_i}{\partial x_i}] \tag{3}$$

The Standard *k-ε* turbulance model covers various applications of turbulence problems [37]. Therefore, the equations for k and ε are as follows:

$$\frac{\partial}{\partial_{xj}}(\rho\, k u_i) = \frac{\partial}{\partial_{xj}}[(\alpha_k \mu_{eff}) \frac{\partial k}{\partial_{xj}}] + G_k - \rho \varepsilon \tag{4}$$

$$\frac{\partial}{\partial_{xj}}(\rho \varepsilon u_i) = \frac{\partial}{\partial_{xj}}[(\alpha_\varepsilon \mu_{eff}) \frac{\partial \varepsilon}{\partial_{xj}}] + C_{1\varepsilon} \frac{\varepsilon}{k}(G_k) - C_{2\varepsilon} \rho \frac{\varepsilon^2}{K} \tag{5}$$

For more details about the employed turbulence model, please refer to the relevant works [20, 38, and 39].

2.4. Mesh Independence Analysis

The density of grids should be higher in areas that contain larger gradients or where important events occur in those areas. According to the flow regime in the area near the moving vanes, which is resolved rotationally, the occurrence of the vortex phenomenon was predicted in this area, and finer grids have been used. Grid sensitivity by successively fine-tuning the grid in the computing area has led to a solution with no dependence on finer grids. On the other hand, the excessive increase in the number of grids increases the calculation time significantly. Therefore, it is necessary to make a compromise between the number of grids and the calculation time. So, to reach the appropriate grid density for the desired solution, this problem has been investigated with different grid densities. A picture of the generated 2D grid is demonstrated in Figure 2. The results obtained from this investigation are shown in Figure 3. Based on the results of this table, the grid with a number of 1100,000 cells has been selected as the appropriate grid to solve this problem.

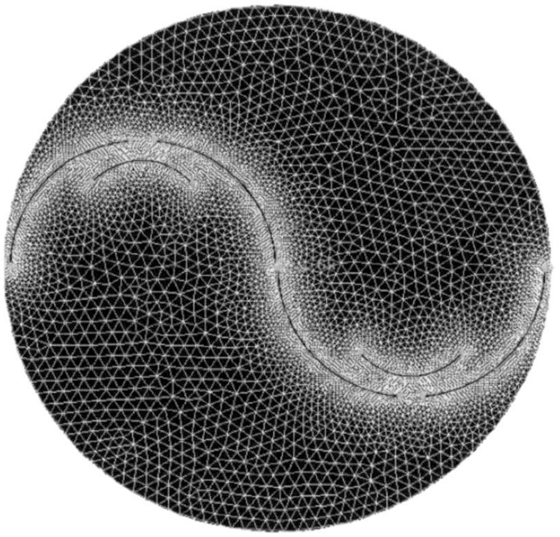

Figure 2. A picture of the generated 2D grid

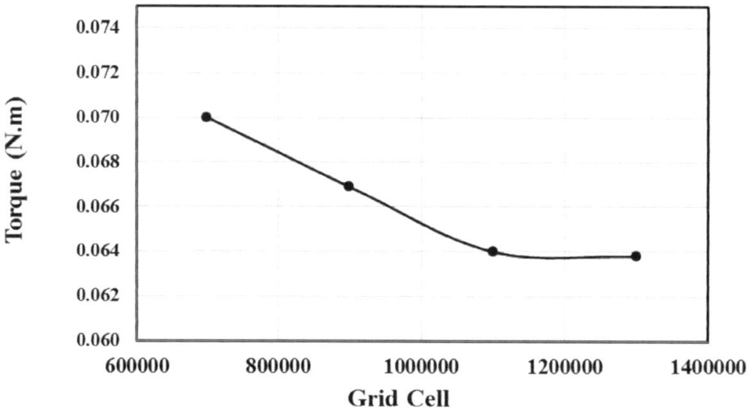

Figure 3. Mesh sensitivity analysis for the present study

2.5. Verification of the Numerical Model

To document the numerical solution, first, the numerical method has been compared with the work of Rodrigues et al. [40] to determine the appropriate numerical solution method for the present problem. For this purpose, a three-bladed Saunius rotor was selected and the output torque coefficient from it at different angles was solved and investigated with all three *k-ε* turbulence models. As can be seen in Figure 4, the *RNG* model is better than the other two models, and the reason for this can be considered to be the rotation of the investigated flow. In the second part of documenting the numerical solution, to check the accuracy of the results obtained from the solution, the outputs have been compared with the experimental results obtained from the work of Mohammad Hadi Ali [31]. As can be seen in Figure 5, the obtained results have a low error with the experimental results, which shows the correctness of the information obtained from the numerical solution.

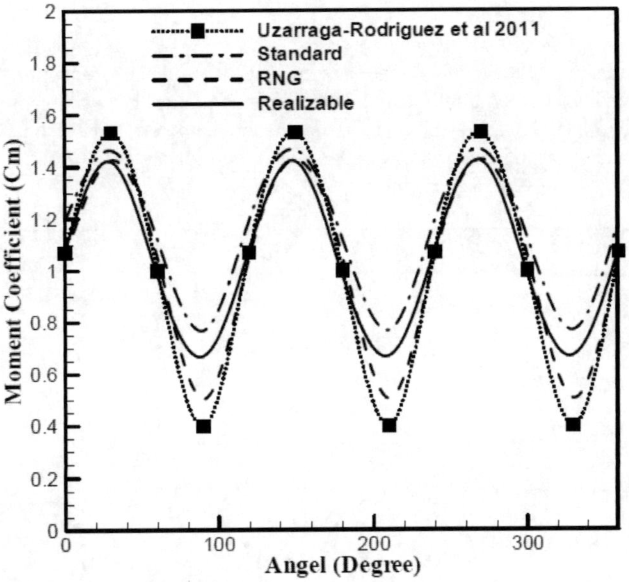

Figure 4. Investigating the 3D numerical solution of three-blades geometry by Rodrigues et al. [40]

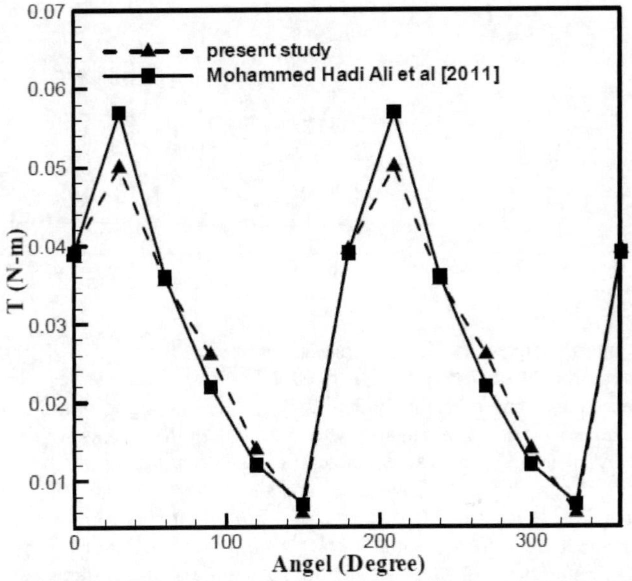

Figure 5. Comparing the 2D numerical solution with the experimental results of Mohammad Hadi Ali [31]

3. Analyzing Results

In this part, the results obtained from the numerical analysis of the desired Savonius turbine equipped with the second blade are presented. The impacts of two geometric parameters of the secondary blades, including the height of secondary blades (h) and the distance between the main and the secondary blades (b) on the output torque were analyzed.

3.1. Impact of the Second Blade Height

In this part, the influence of the height of the second blade (h) on the production torque of the proposed Savonius turbine has been investigated numerically. Three different values have been considered for the height of the second blade, including h = 3, 6, and 12 mm. The value of the diameter of the second blade is fixed and equal to 12 mm, and all the simulations of this part have been performed for the case of b = 0 mm. The schematics of the examined models here are illustrated in Figure 6.

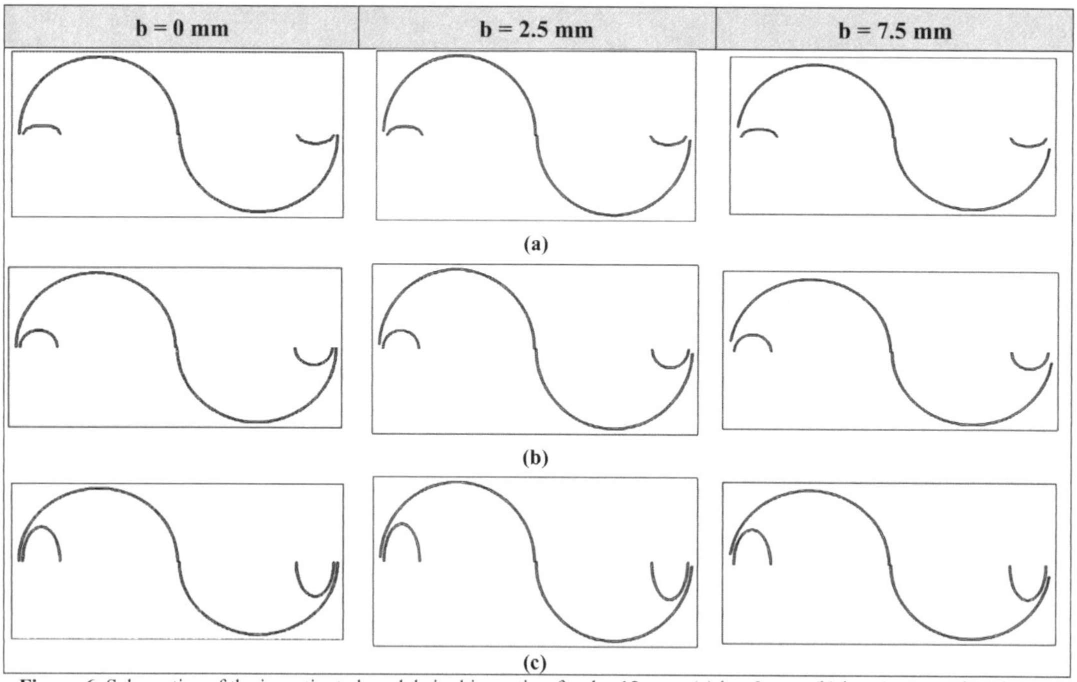

Figure 6. Schematics of the investigated models in this section for d = 12 mm; (a) h = 3 mm, (b) h = 6 mm, and (c) h = 12 mm

The value of the produced torque for different angular positions of the Savonius turbine blade and various values of the height of the second blade (h) are demonstrated in Figure 7. As can be seen, in this case, the difference between the models is very small, however, the model h = 3 mm has the highest production torque, albeit with a small difference among the other models. It can also be seen that with the increase of the value of h, the position of the maximum torque is also shifted and as an example, it can be seen that for the models with the height of the second blade of h = 6 and 12 mm, the peak points are at θ = 330 degrees (Figure 7a). Accorsing to the Figure 7b (b = 2.5 mm), as the second blade height increases from h = 3 mm to h = 12 mm (300% growth), the maximum produced torque rises by about 25%. Also, Figure 7c (b = 7.5 mm) depicts that by augmenting the second blade height by 300% (from h = 3 mm to h = 12 mm), the maximum produced torque increases by about 60%.

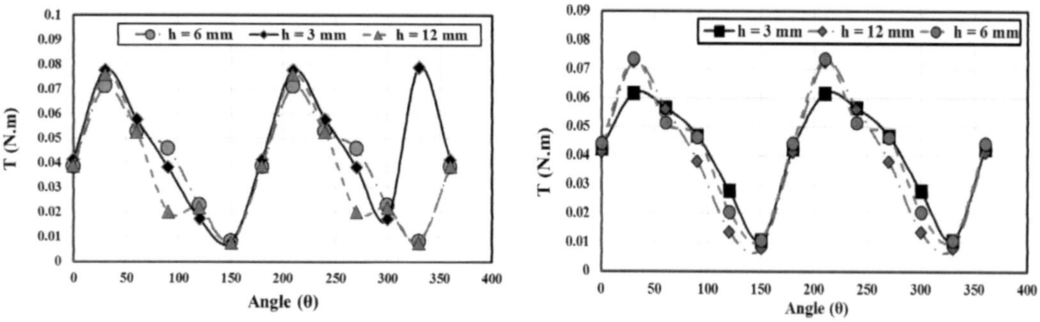

(a) (b)

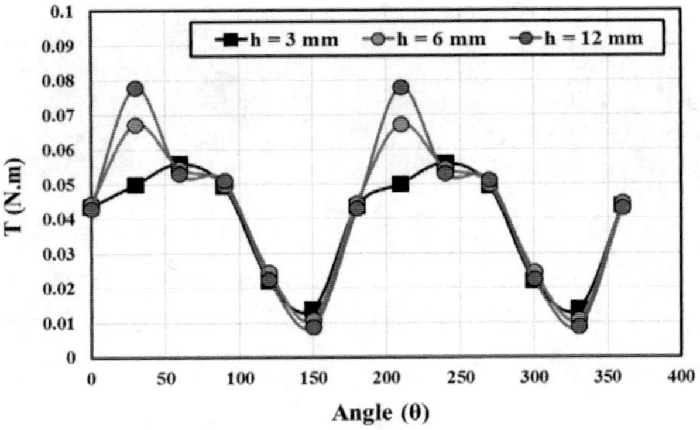

(c)

Figure 7. Production torque of the proposed Savonius turbine in the case of d = 12 mm for (a) b = 0 mm, (b) b = 2.5 mm, and (c) b = 7.5 mm

The pressure contour for different models in the case of d = 12 mm and b = 0 mm is shown in Figure 8. To better see the differences and the effect of h, only the results for two values of h = 3 and 12 mm are shown. As can be seen, the difference between the contours, similar to the results shown in Figure 7, is small, but still, it can be compared that the negative pressure (blue area) behind the main vane in the case of h = 12 mm is slightly It is more than the case of h = 3 mm, which itself becomes a factor for reducing the produced torque. For a better understanding of the results, the contour of the velocity value for different models in the case of d = 12 mm and b = 0 mm is shown in Figure 9. As can be seen, the difference between the contours is small, similar to the results shown in Figures 7 and 8, but still it can be compared that the blue area, which indicates zero velocity and the existence The return flow behind the main vane in the case of h = 12 mm is slightly higher than the case of h = 3 mm, which is a factor in reducing the produced torque.

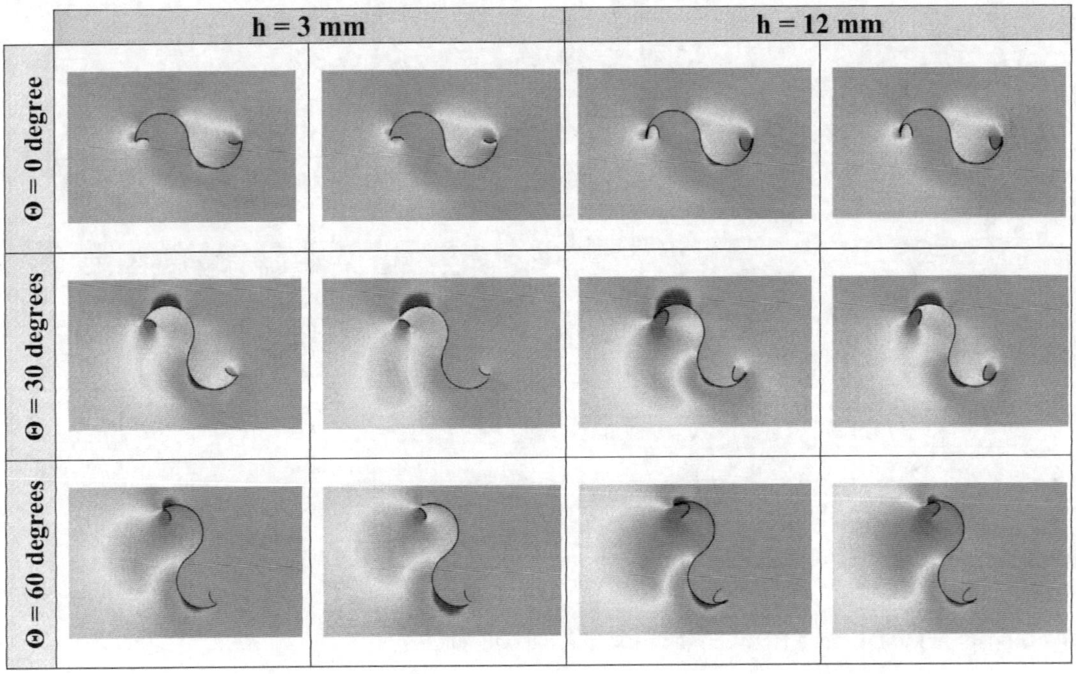

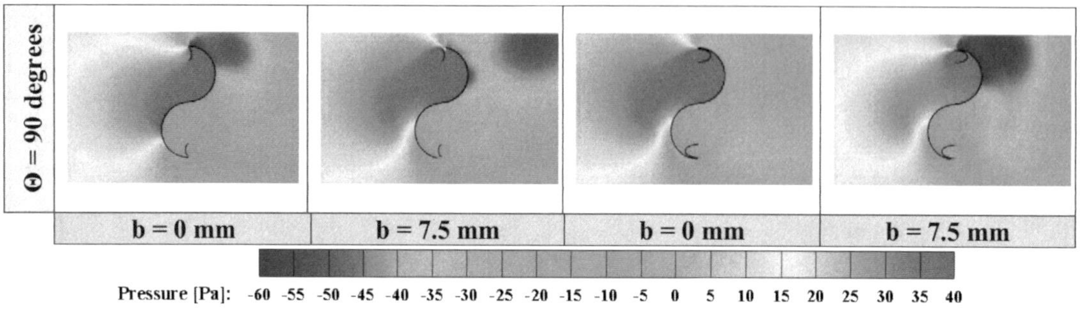

Figure 8. Pressure contours for two different models (b = 0 and 7.5 mm) in d=12 mm

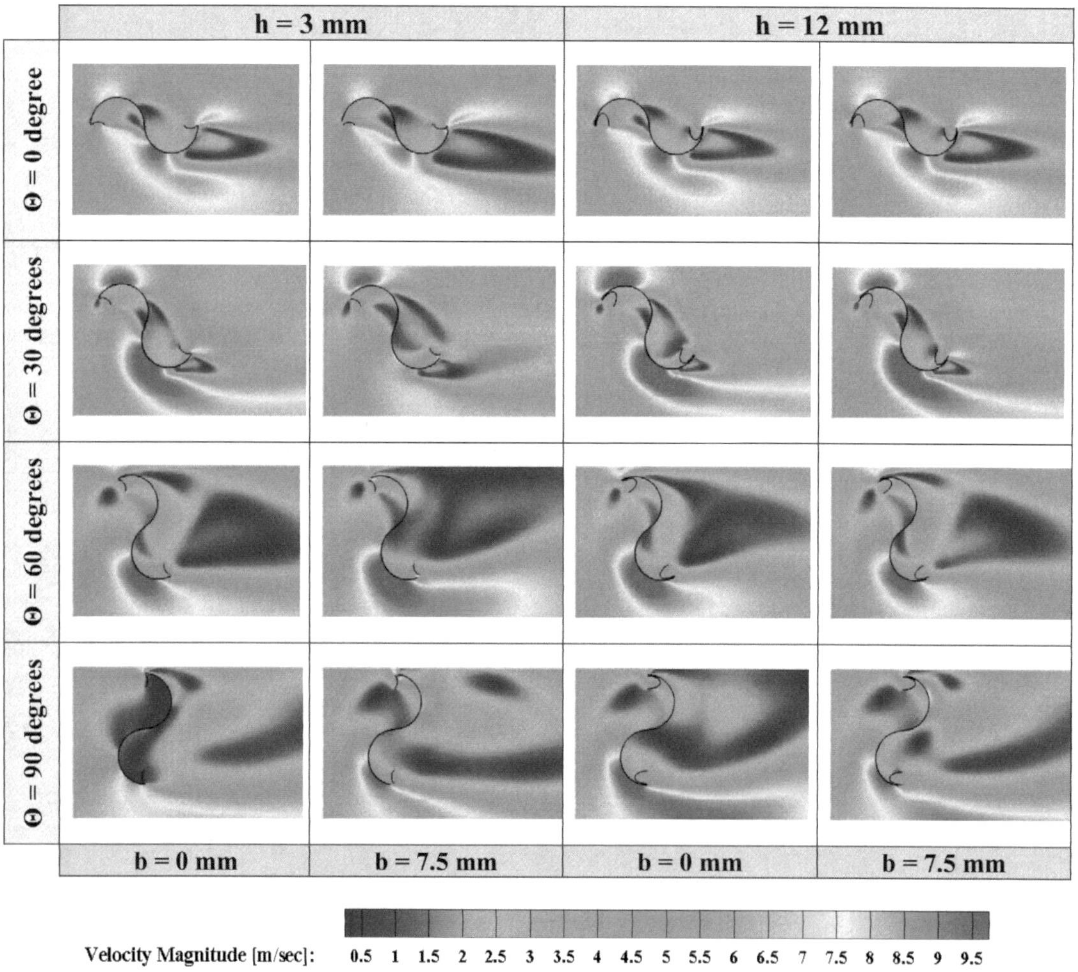

Figure 9. Contours of velocity value for two different models (b = 0 and 7.5 mm) in the case of d=12 mm

3.2. Impact of the Distance Between the First and Second Blades

In this part, the effect of the distance between the first and second blades on the torque produced by the Savonius turbine has been investigated. Three different values have been considered for the distance between the first and second blades, including zero, 2.5, and 7.5 mm. The diameter of the second blade is constant and equal to 12 mm and h = 6 mm. The schematic of the models examined in this section is shown in Figure 10. As can be seen, three different values have been considered for the distance of the second blade from the main blade, including zero, 2.5, and 7.5 mm. The value of the produced torque for different placement angles of the Savonius turbine blade and different values of the distance of the second blade from the main blade is shown in Figure 11. As can be seen, firstly, the difference between the examined models is small and compared to the case of h=3 mm, it is very small in this case. Secondly, it can be seen that in this case (h = 6 mm), the highest production torque belongs to the b = 2.5 mm model.

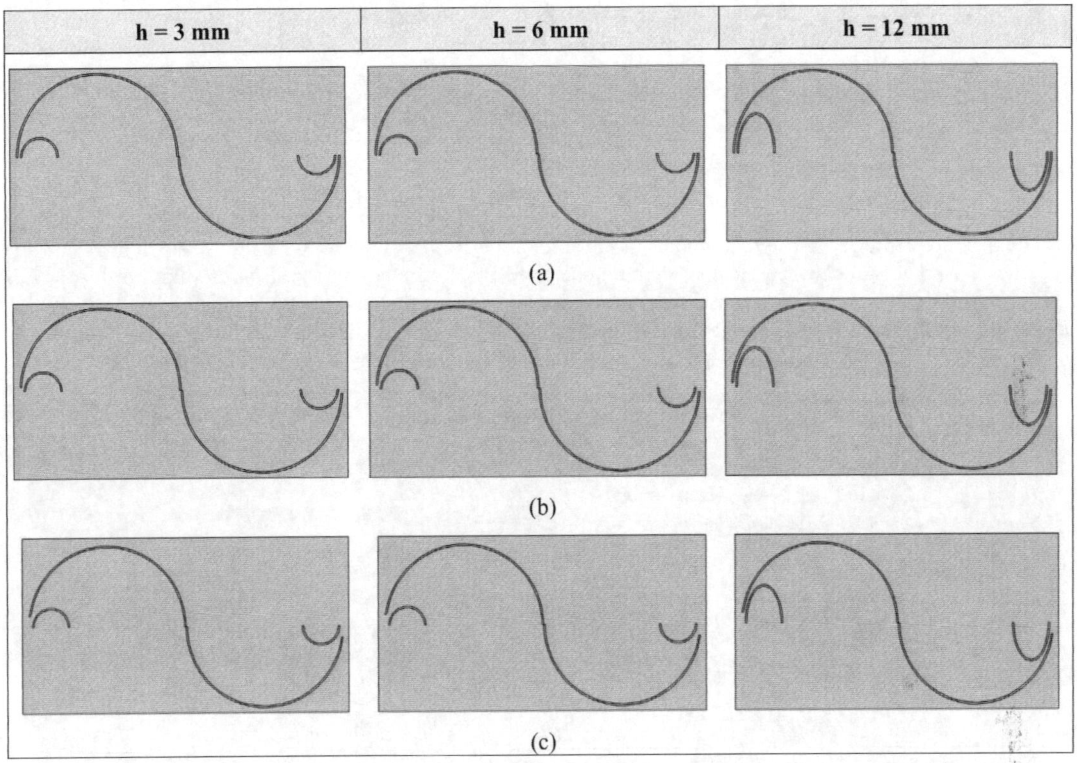

Figure 10. Schematics of the investigated models in this section for d = 12 mm; (a) b = 0 mm, (b) b = 2.5 mm, and (c) b = 7.5 mm

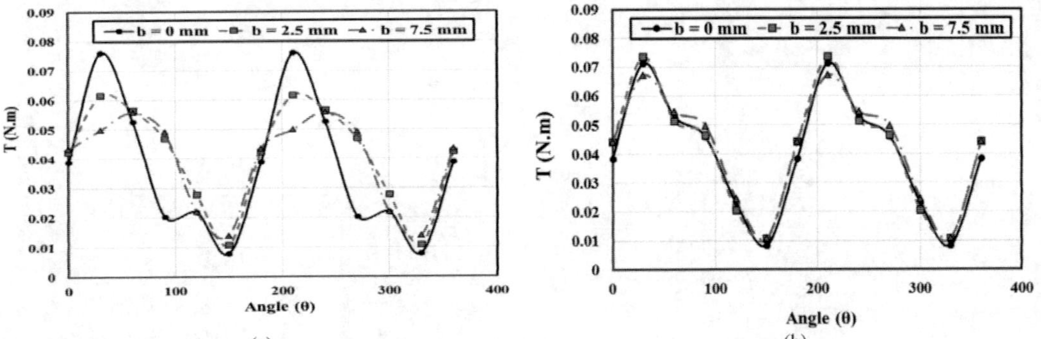

(a) (b)

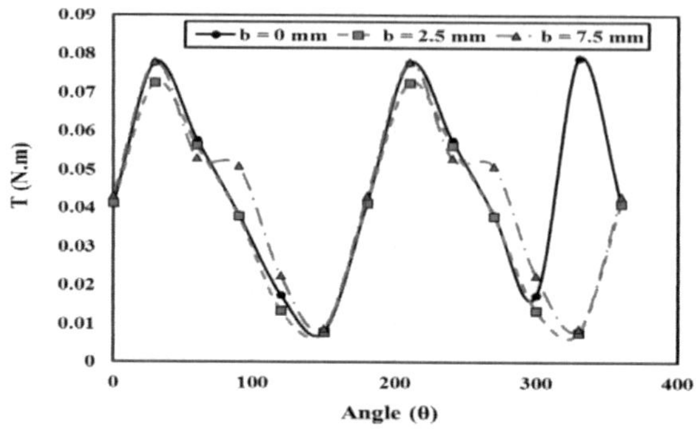

(c)

Figure 11. Production torque of desired Savonius turbine in the case of d=12 mm for (a) h = 3 mm, (b) h = 6 mm, and (c) h = 12 mm

The pressure contour for different models in the case of d = 12 mm and h = 3-12 mm is shown in Figure 12. For a better comparison between the considered models, the results are shown for all three values of b = 0, 2.5, and 7.5 mm. As can be seen, in this case, there is an optimal point, so that in the b = 2.5 mm model, the blue area behind the vanes, which indicates negative pressure, is less, which leads to more productive torque. compared to other models. The difference between the other two models is small and cannot be seen in the pressure contour.

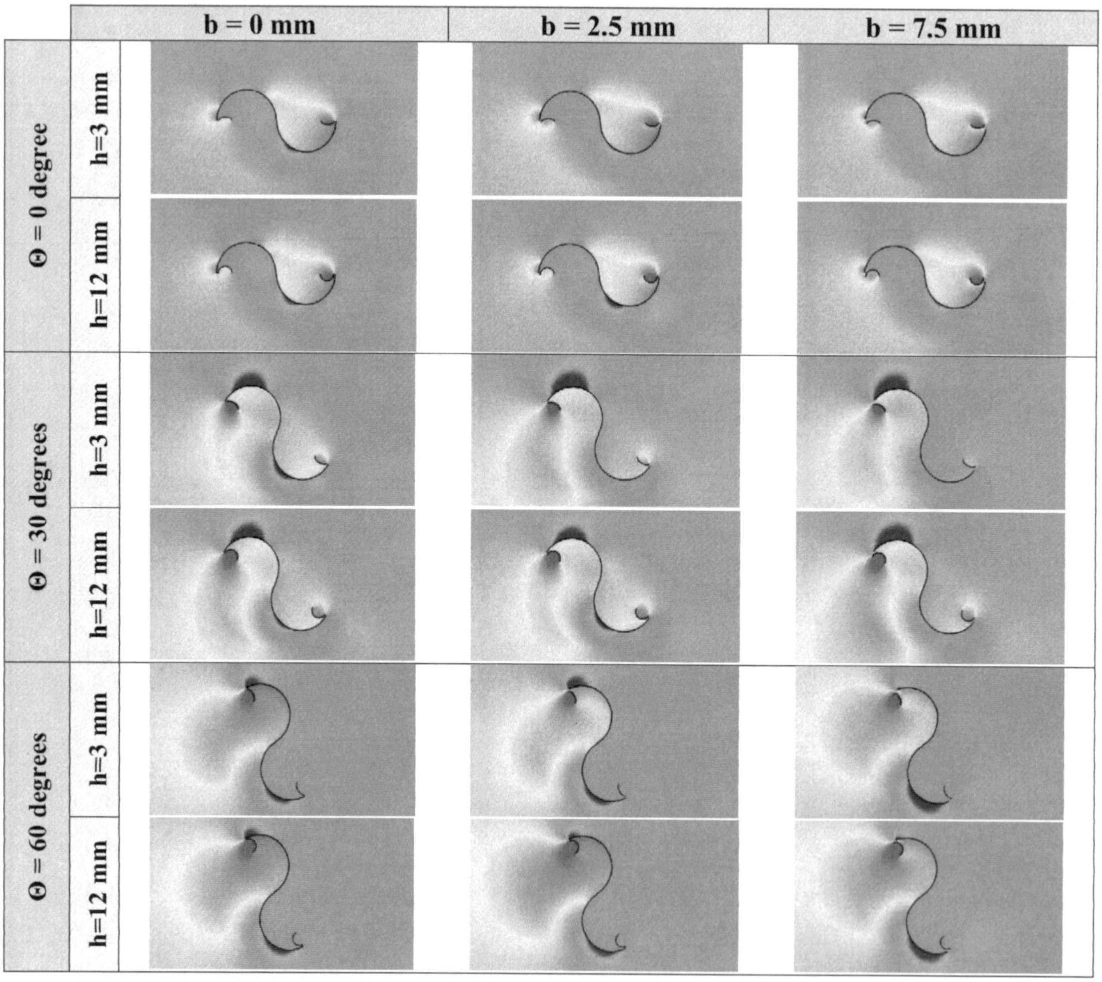

IOP Conf. Series: Earth and Environmental Science 1149 (2023) 012006 doi:10.1088/1755-1315/1149/1/012006

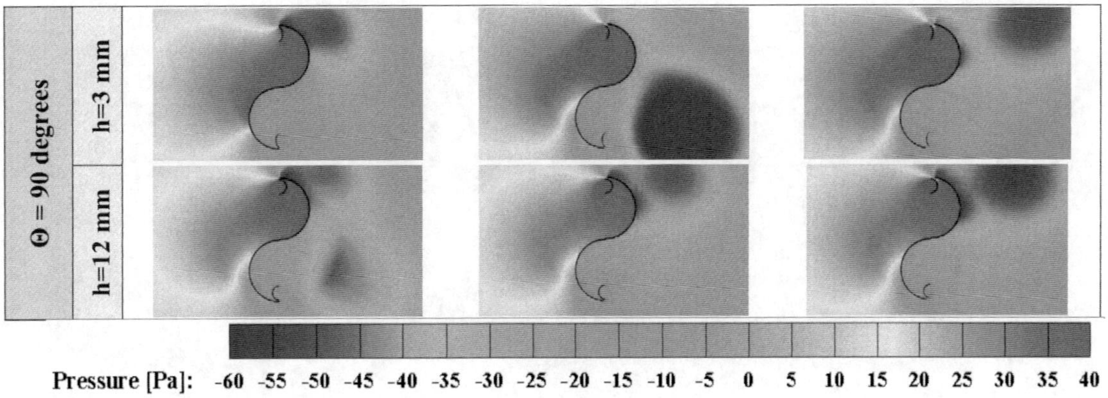

Figure 12. Pressure contours for different models in d = 12 mm

For a better understanding of the results, the contour of the velocity value for different models in the case of d = 12 mm and h = 3-12 mm is shown in Figure 13. As can be seen, according to the results shown in Figures 12 and 13, there is an optimal point, so that in the model b = 2.5 mm, the blue area behind the blades which represents The rotation and return currents are less, which leads to more productive torque than other models. The difference between the other two models is small and the same velocity value as the pressure cannot be seen in the contour.

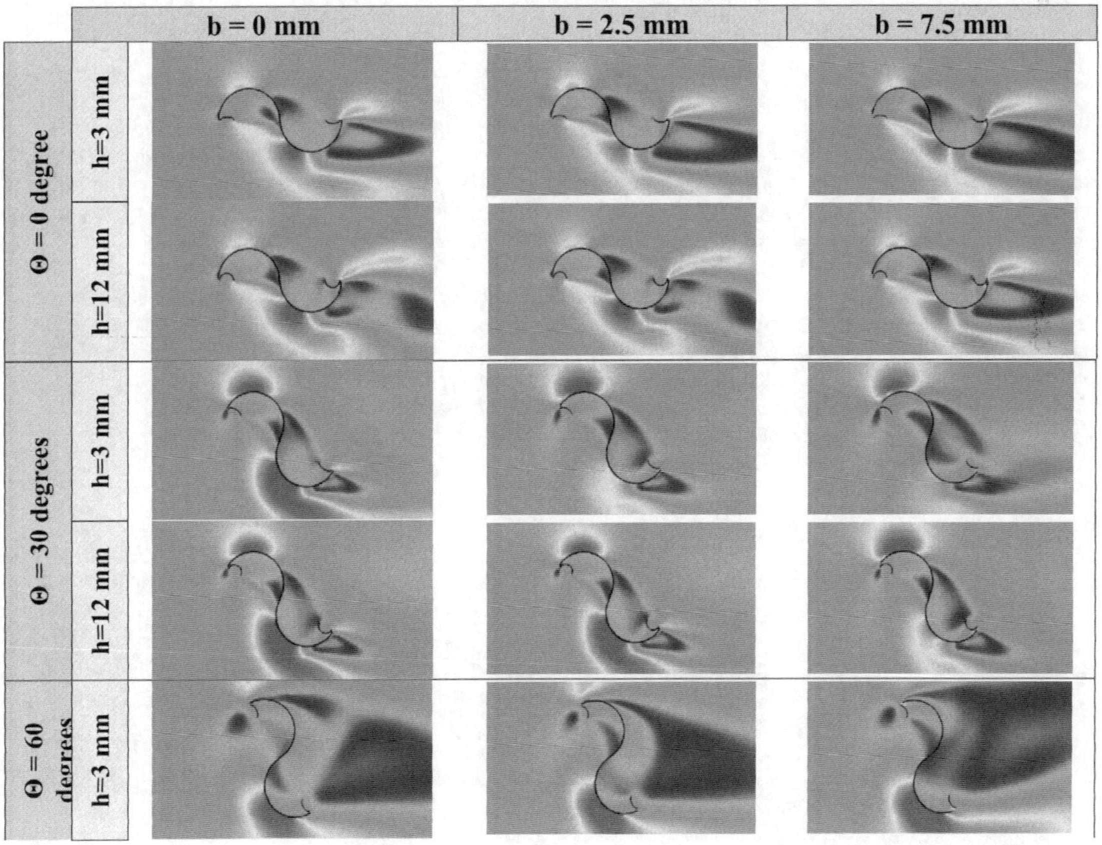

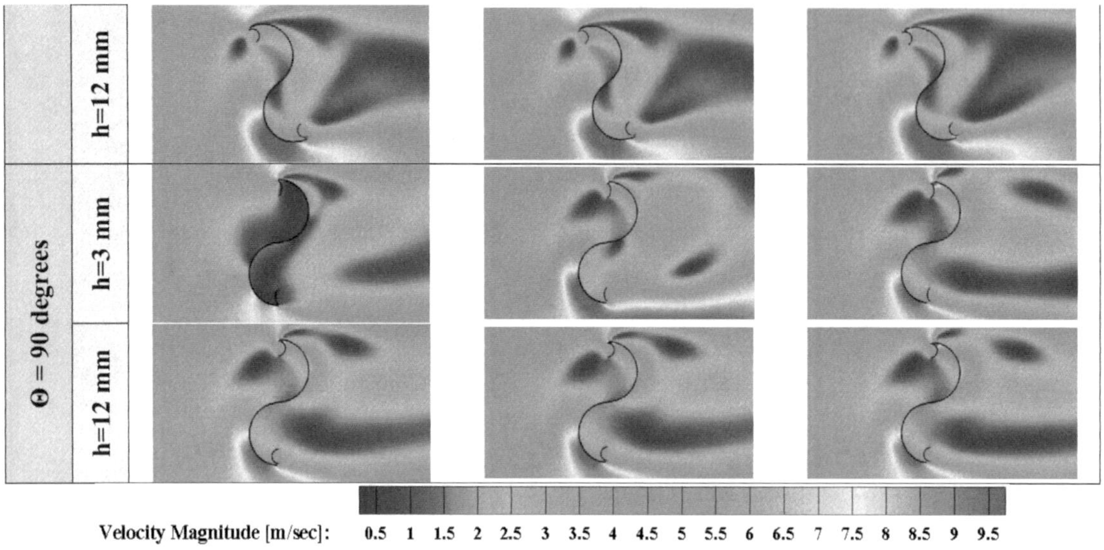

Figure 13. Velocity magnitude contours for different models in d = 12 mm

4. Conclusions

In this work, the performance of the Savonius vertical axis wind turbine with an additional blade was numerically investigated and the effect of two geometrical parameters, the height of the second blade and the distance between the main blade and the second blade, on the produced torque was evaluated and analyzed. Numerical simulations were carried out by ANSYS FLUENT 18.2 commercial software. In addition to the produced torque, the obtained numerical results are also presented in the form of pressure contours and velocity values. Numerical results have shown that the use of the second blade has a positive effect on the production torque compared to the case without the second blade. The geometric parameters of the second blade have a significant effect on the performance of the proposed turbine, and the choice of these items has a great impact on the performance of the proposed turbine. For the case b = 2.5 mm, as the second blade height increases from h = 3 mm to h = 12 mm (300% growth), the maximum produced torque rises by about 25%. Also, for the case b = 7.5 mm, by augmenting the second blade height by 300% (from h = 3 mm to h = 12 mm), the maximum produced torque increases by about 60%. Furthermore, by increasing the distance between the secondary and the main blades from b = 0 mm to b = 2.5 mm and b = 7.5 mm, the amount of torque produced decreases by 17.33 and 26.66%, respectively. Moreover, by augmenting the distance between the secondary and main blades from b = 2.5 mm to b = 7.5 mm (200% growth), the produced torque decreases by 11.3%.

References

[1]. Ajarostaghi, S. S. M., & Mousavi, S. S. (2022). Solar energy conversion technologies: Principles and advancements. In *Solar Energy Advancements in Agriculture and Food Production Systems* (pp. 29-76). Academic Press.

[2]. Nouri Kadijani, O., Kazemi Moghadam, H., Mousavi Ajarostaghi, S. S., Asadi, A., & Saffari Pour, M. (2022). Hydrothermal performance of humid air flow in a rectangular solar air heater equipped with V-shaped ribs. *Energy Science & Engineering*.

[3]. Olfian, H., Ajarostaghi, S. S. M., Ebrahimnataj, M., Farhadi, M., & Arıcı, M. (2022). On the thermal performance of evacuated tube solar collector integrated with phase change material. *Sustainable Energy Technologies and Assessments*, *53*, 102437.

[4]. Liu, H., Li, Y., Duan, Z., & Chen, C. (2020). A review on multi-objective optimization framework in wind energy forecasting techniques and applications. *Energy Conversion and Management*, 224, 113324.

[5]. Mousavi Ajarostaghi, S. S., Javadi, H., Mousavi, S. S., Poncet, S., & Pourfallah, M. (2021). Thermal performance of a single U-tube ground heat exchanger: A parametric study. *Journal of Central South University*, 28(11), 3580-3598.

[6]. Javadi, H., Ajarostaghi, S. S. M., Mousavi, S. S., & Pourfallah, M. (2019). Thermal analysis of a triple helix ground heat exchanger using numerical simulation and multiple linear regression. *Geothermics*, 81, 53-73.

[7]. Javadi, H., Mousavi Ajarostaghi, S. S., Rosen, M. A., & Pourfallah, M. (2018). A comprehensive review of backfill materials and their effects on ground heat exchanger performance. *Sustainability*, 10(12), 4486.

[8]. Hand, B., Kelly, G., & Cashman, A. (2021). Aerodynamic design and performance parameters of a lift-type vertical axis wind turbine: A comprehensive review. *Renewable and Sustainable Energy Reviews*, 139, 110699.

[9]. Spera, D. A. (1994). *Wind turbine technology*.

[10]. Tong, W. (2010). *Wind power generation and wind turbine design*. WIT press.

[11]. Johari, M. K., Jalil, M., & Shariff, M. F. M. (2018). Comparison of horizontal axis wind turbine (HAWT) and vertical axis wind turbine (VAWT). *International Journal of Engineering and Technology*, 7(4.13), 74-80.

[12]. Porté-Agel, F., Bastankhah, M., & Shamsoddin, S. (2020). Wind-turbine and wind-farm flows: a review. *Boundary-Layer Meteorology*, 174(1), 1-59.

[13]. Du, Y., Zhou, S., Jing, X., Peng, Y., Wu, H., & Kwok, N. (2020). Damage detection techniques for wind turbine blades: A review. *Mechanical Systems and Signal Processing*, 141, 106445.

[14]. Tasneem, Z., Al Noman, A., Das, S. K., Saha, D. K., Islam, M. R., Ali, M. F., ... & Alam, F. (2020). An analytical review on the evaluation of wind resource and wind turbine for urban application: Prospect and challenges. *Developments in the Built Environment*, 4, 100033.

[15]. Rezamand, M., Kordestani, M., Carriveau, R., Ting, D. S. K., Orchard, M. E., & Saif, M. (2020). Critical wind turbine components prognostics: A comprehensive review. *IEEE Transactions on Instrumentation and Measurement*, 69(12), 9306-9328.

[16]. Modi, V. J., & Fernando, M. S. U. K. (1989). *On the performance of the Savonius wind turbine*.

[17]. Zemamou, M., Aggour, M., & Toumi, A. (2017). Review of savonius wind turbine design and performance. *Energy Procedia*, 141, 383-388.

[18]. Abraham, J. P., Plourde, B. D., Mowry, G. S., Minkowycz, W. J., & Sparrow, E. M. (2012). Summary of Savonius wind turbine development and future applications for small-scale power generation. *Journal of Renewable and Sustainable Energy*, 4(4), 042703.

[19]. Longo, R., Nicastro, P., Natalini, M., Schito, P., Mereu, R., & Parente, A. (2020). Impact of urban environment on Savonius wind turbine performance: A numerical perspective. *Renewable energy*, 156, 407-422.

[20]. Norouztabar, R., Mousavi Ajarostaghi, S. S., Mousavi, S. S., Nejat, P., Rahimian Koloor, S. S., & Eldessouki, M. (2022). On the performance of a modified triple stack blade Savonius wind turbine as a function of geometrical parameters. *Sustainability*, 14(16), 9816.

[21]. Honarbari, A., Najafi-Shad, S., Saffari Pour, M., Ajarostaghi, S. S. M., & Hassannia, A. (2021). MPPT Improvement for PMSG-Based Wind Turbines Using Extended Kalman Filter and Fuzzy Control System. *Energies*, 14(22), 7503.

[22]. Ferrari, G., Federici, D., Schito, P., Inzoli, F., & Mereu, R. (2017). CFD study of Savonius wind turbine: 3D model validation and parametric analysis. *Renewable energy*, 105, 722-734.

[23]. Altan, B. D., & Atılgan, M. (2008). An experimental and numerical study on the improvement of the performance of Savonius wind rotor. *Energy Conversion and Management*, *49*(12), 3425-3432.

[24]. Mohamed, M. H., Janiga, G., Pap, E., & Thévenin, D. (2011). Optimal blade shape of a modified Savonius turbine using an obstacle shielding the returning blade. *Energy Conversion and Management*, *52*(1), 236-242.

[25]. Kamoji, M. A., Kedare, S. B., & Prabhu, S. V. (2009). Experimental investigations on single stage modified Savonius rotor. *Applied Energy*, *86*(7-8), 1064-1073.

[26]. Hayashi, T., Li, Y., & Hara, Y. (2005). Wind tunnel tests on a different phase three-stage Savonius rotor. *JSME International Journal Series B Fluids and Thermal Engineering*, *48*(1), 9-16.

[27]. Saha, U. K., Thotla, S., & Maity, D. (2008). Optimum design configuration of Savonius rotor through wind tunnel experiments. *Journal of Wind Engineering and Industrial Aerodynamics*, *96*(8-9), 1359-1375.

[28]. Fujisawa, N., Ishimatsu, K., & Kage, K. (1995). A comparative study of Navier-Stokes calculations and experiments for the Savonius rotor.

[29]. Kawamura, T., Hayashi, T., & Miyashita, K. (1998). 41. Application of the Domain Decomposition Method to the Flow around the Savonius Rotor. In *Domain Decomposition Methods in Sciences and Engineering* (p. 393).
[30]. Gupta, R., Debnath, B., & Das, R. (2009). CFD analysis of a two-bucket Savonius rotor using Fluent package. In *Proc. of the European Wind Energy Conference*.
[31]. Ali, M. H. (2013). Experimental comparison study for Savonius wind turbine of two & three blades at low wind velocity. *International Journal of Modern Engineering Research (IJMER)*, *3*(5), 2978-2986.
[32]. Deb, B., & Gupta, R. (2012). Fluid flow analysis of Savonius rotor at different rotor angle using CFD. *ISESCO JOURNAL of Science and Technology*, *8*(14), 35-42.
[33]. Yoshida, Y., & Kawamura, T. (2012). Numerical simulations of two-dimensional flows around a Savonius rotor with various curvature of the blade. *お茶の水女子大学自然科学報告*, *63*(1), 11-21.
[34]. McTavish, S., Feszty, D., & Sankar, T. (2012). Steady and rotating computational fluid dynamics simulations of a novel vertical axis wind turbine for small-scale power generation. *Renewable energy*, *41*, 171-179.
[35]. Antar, E., & Elkhoury, M. (2019). Parametric sizing optimization process of a casing for a Savonius Vertical Axis Wind Turbine. *Renewable Energy*, 136, 127-138.
[36]. Ahmed, W. U., Zahed, M. J. H., Rahman, M. A., & Mamun, M. (2014, September). Numerical study of two and three bladed Savonius wind turbine. In *2nd International Conference on Green Energy and Technology* (pp. 36-40). IEEE.
[37]. *ANSYS, F. 18.2 User Guide*. Knowledge Resources: Online Documentation.
[38]. Saedodin, S., Zaboli, M., & Ajarostaghi, S. S. M. (2021). Hydrothermal analysis of heat transfer and thermal performance characteristics in a parabolic trough solar collector with Turbulence-Inducing elements. *Sustainable Energy Technologies and Assessments*, 46, 101266.
[39]. Ajarostaghi, S. S. M., Zaboli, M., & Nourbakhsh, M. (2021). Numerical evaluation of turbulence heat transfer and fluid flow of hybrid nanofluids in a pipe with innovative vortex generator. *Journal of Thermal Analysis and Calorimetry*, 143(2), 1583-1597.
[40]. Uzarraga-Rodriguez, N. C., Gallegos-Mun˜ oz, A., & Riesco A´ vila, J. M. (2011, January). Numerical analysis of a rooftop vertical axis wind turbine. In *Energy Sustainability* (Vol. 54686, pp. 2061-2070).

Probabilistic Analysis of RC Buildings Based on Incremental Dynamic Analysis

Merin Mathews[1], **B R Jayalekshmi**[2], **Katta Venkataramana**[2]

[1] Research Scholar, Department of Civil Engineering, NITK, Surathkal, Mangaluru, India
[2] Professor, Department of Civil Engineering, NITK, Surathkal, Mangaluru, India

Abstract. Earthquake is a natural disaster that can induce immense damage to properties and human lives. Fragility curve, defined as the probability of reaching or exceeding a specific damage state under an earthquake excitation, provides a prediction of damage that may occur during an earthquake. Incremental Dynamic Analysis Curve (IDA curve) developed from Incremental Dynamic analysis can be considered as the initial step of fragility curve development. The present study summarises the development of IDA curve for a RC building considering maximum roof displacement and storey drift as Damage Measure (DM) and peak ground acceleration as Intensity Measure (IM). The fragility curves developed using incremental dynamic analysis provide an overview of the probability of exceedance of damage limits in an existing building when acted upon by different levels of earthquake excitations and the performance level can be improved by incorporating appropriate design changes.
Keywords: fragility curve; incremental dynamic analysis; IDA curve; damage measure; intensity measure.

1. Introduction

The approach of engineering design is to design structures in such a way that they can survive under the most severe earthquake, during their service lives and to minimize the loss of life and the probability of damage. Even though we cannot prevent the earthquakes from occurring, we can develop seismic risk mitigation methods based on tools for probabilistic assessment of performance of structures. Fragility curve, defined as the probability of reaching or exceeding a specific damage state under an earthquake excitation, provide an estimate of possible damage during occurrence of an earthquake. Engineers need to focus on development of probabilistic risk models that can be effectively used for quantification of damage.

Many studies have been conducted on the development of fragility curve considering the damage state indicating the quantification of damage of the structure. The different damage states were fixed as slight, moderate and collapse level considering different occupancy levels as immediate occupancy, life safety and the collapse prevention. The damage limit corresponding to each damage state will vary depending upon the type of structure chosen [1]. Fragility curves provide an overview of expected seismic performance and can be utilized in deciding the retrofitting methods, estimating financial losses, and recommending appropriate changes of building codes to ensure better performance of structures [2].

1.1. Incremental Dynamic Analysis

Incremental Dynamic Analysis (IDA) method is a valuable tool for parametric analysis that can address both demand and capacity of structures. Incremental Dynamic Analysis Curve (IDA curve) developed from Incremental Dynamic analysis can be considered as the initial step of fragility curve development.

In this method a structural modal is subjected to a set of ground motions scaled to different levels and for each set of ground motion IDA curves of chosen response parameters are plotted for different intensity levels [3]. The expected response and the extent of damage of structures under different earthquake excitation intensities can be analysed using Incremental dynamic analysis. This method provides a more accurate and realistic result about the response of buildings under seismic excitations of different intensities when compared to static analysis methods. The IDA curves can be developed based on the relationship between an Intensity Measure and Damage Measure [4].

Fragility curves developed using Incremental Dynamic Analysis can be used to investigate the extent of seismic effect on structures and to predict the possible damages under different seismic excitations. To assess the extent of damage, different damage state levels can be set for structures [5].

The present study summarises the development of IDA curve for a RC building considering maximum roof displacement and storey drift as Damage Measure (DM). Peak Ground Acceleration is selected as Intensity Measure (IM) to estimate the intensity of ground motion where a structure reaches or exceeds a particular performance level [6]. The structural model is applied with a set of five earthquake excitations scaled to different levels and for each set of ground motion IDA curves of chosen response parameters are plotted for different intensity levels.

1.2. Structure Model

A G+3 storey RC moment resisting frame in seismic zone V (very severe zone) founded on soil type II (medium stiff soil) with plan dimension of 12m x 16m is selected as the model for analysis. Concrete of M25 grade and Fe415 steel have been assumed.

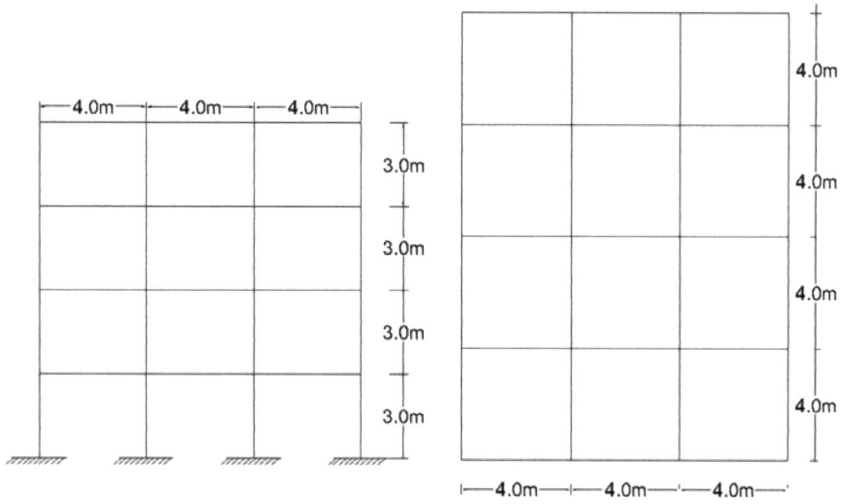

Figure 1. (a) Elevation; (b) Plan

1.3. Selection of Ground Motion

Selection of ground motion plays an important factor in estimating the response of structures in the non-linear analysis [1]. The uncertainty in the selection of ground motion for incremental dynamic analysis can be minimized by selecting many ground motions from different regions and scaling of each selected ground motion is carried out to match the design spectrum for which the structure is modelled. In this study the ground motions are scaled to match the design spectrum corresponding to Zone V and Type II soil. Five sets of earthquake data are selected for this example and are scaled to 0.1g, .02g, 0.3g, 0.4g up to 1g. The details of the earthquake selected are listed in table 1.

Table 1. Characteristics of ground motions

Event	Station	PGA(g)	Magnitude	Notation
Big bear 1992	North Shore	0.043	6.46	EQ1
Imperial Valley 1979	El Centro E10	0.176	6.53	EQ2
Kobe 1995	Kakogawa	0.058	6.9	EQ3
Landers 1992	Mission Creek Fault	0.122	7.28	EQ4
Northridge 1994	Arleta	0.237	6.69	EQ5

Incremental Dynamic analysis was conducted in ETABS. IDA curves were plotted with PGA of the scaled time history ground motions as the Intensity Measure [4]. Roof Displacement and Storey Drift (%) were taken as the damage measure for developing IDA curves.

2. Results and Discussions

The accuracy and reliability of the result depend upon defining the correct relationship between damage level and damage status of different structures considered. Based on the difference in age, number of stories and design methods different damage levels are obtained at different drift levels for each type of building. To quantify the damage of the structure using fragility curve, damage state levels were fixed as Slight, Moderate, Extensive and Collapse Damage Level [1, 7]. Damage state levels based on drift ratio are given in Table 2.

Table 2. Damage State Levels

Damage State	Collapse	Extensive	Moderate	Slight
Drift Ratio	4%	1.6%	0.64%	0.4%

Fig. 2 shows the IDA curves developed for roof displacement using incremental time history analysis. For all five sets of earthquake excitations, when the PGA values were low, the structure did not undergo significant damage and as the PGA values increase there were considerable increase in the roof displacement. Using IDA curve, depending upon the type of structure we will be able estimate the intensity at which the structure reaches the safe limit of displacement.

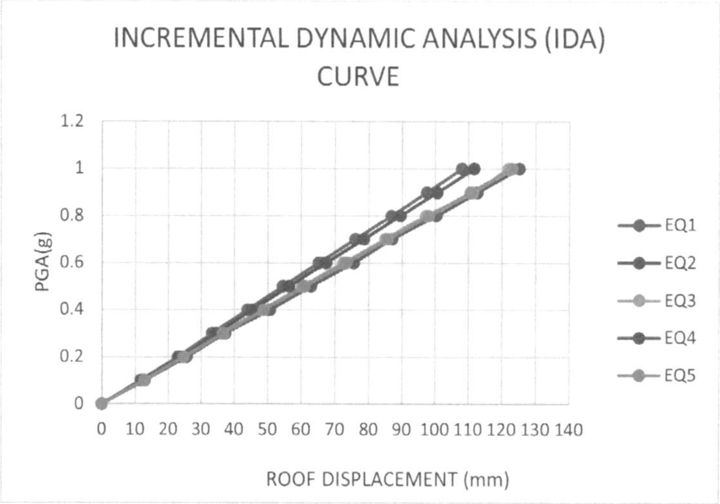

Figure 2. Incremental Dynamic Analysis (IDA) Curve for Roof Displacement

Fig. 3 shows the IDA curves developed for storey drift using incremental time history analysis. As in the case of IDA curve for roof displacement, for earthquake excitations with PGA up to 0.3g the damage levels were within the safe limit of Immediate Occupancy Level. Beyond a PGA value of 0.4g the extent of damage exceeds the slight damage level. By knowing the PGA value beyond which the structure undergoes significant damage, strengthening techniques can be adopted if necessary to avoid damage to people and property [8].

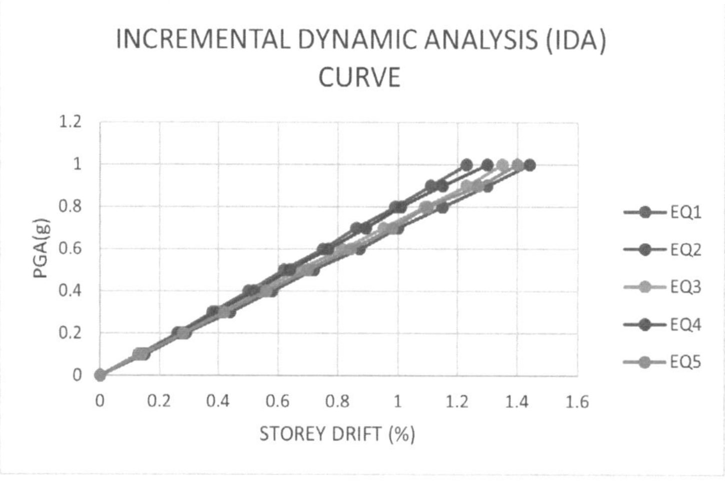

Figure 3. Incremental Dynamic Analysis (IDA) Curve for Storey Drift

Fragility curves are log-normal functions which express the probability of reaching or exceeding a specific damage state [2]. The lognormal Cumulative Distribution Function (CDF) is the most used form of fragility function [9]. It is represented as

$$F_d(x) = \emptyset(\ln(\frac{x}{\theta d})/\beta_d) \tag{1}$$

where
β_d= standard deviation of natural logarithm of PGA, θ_d = median of PGA

The fragility curves indicate the probabilistic distribution of the selected damage levels. Figure 4 shows the fragility curves developed for the selected model.

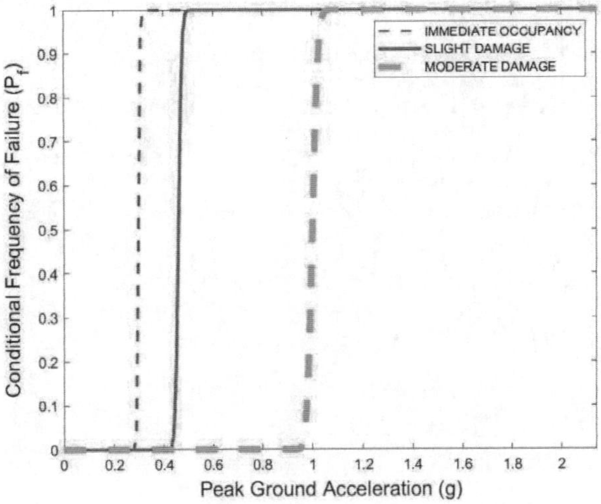

Figure 4. Fragility Curve

From the figure it is clear that when the structure is subjected to ground motions of intensity up to PGA value of 0.4g the probability of reaching slight damage level is zero and as PGA reaches 0.5g, the probability of exceedance of damage limit increases to 100. The probability of occurrence of moderate damage is zero up to ground motion intensity of 0.9g and it increases to 90 % when PGA is 1g. As the structural system is subjected to ground motion with PGA exceeding 1g, the probability of reaching or exceeding the immediate occupancy level as well as moderate damage level reaches to 100%.

3. Concluding Remarks

From the fragility curve it is clear that the probability of exceedance of damage limit is very low when the structure is under weak seismic excitations. As the intensity of excitation increases the probability of exceedance increases and the damage state changes to next level. From this result an estimate on the intensity of earthquake at which a structure reaches or exceeds a particular level of damage can be made. The IDA curves provide an assessment on the damage measure of the structure under different earthquake excitation levels. The fragility curve developed using incremental dynamic analysis provide an overview of the probability of exceedance of damage limits in an existing building when acted upon by different levels of earthquake excitations. Based on the importance and expected life of the building the performance level can be improved by making appropriate design changes incorporating enhancement of ductility and improving the lateral stiffness of the structure. This probabilistic method of

IOP Conf. Series: Earth and Environmental Science 1149 (2023) 012007 doi:10.1088/1755-1315/1149/1/012007

analysis can be used to decide on the retrofitting techniques to be adopted to minimise the extent of damage and financial loss during seismic excitations.

4. References

[1] Roy T. and Agarwal P 2014 Comparison of damage index and fragility curve of RC structure using different Indian standard codes *Structural Engineering Convention IIT Delhi*

[2] Ibrahim E and Mostafa M. El-Shami 2011 Seismic fragility curves for mid-rise reinforced concrete frames in Kingdom of Saudi Arabia *The IES Journal Part A: Civil & Structural Engineering* 213-223

[3 Vamvatsikos D and Cornell A 2002 Incremental dynamic analysis *Earthquake Engineering and Structural Dynamics* **31** 491-514

[4] Kassem M M, Nazri F M and Farsangi E N 2020 On the quantification of collapse margin of a retrofitted university building in Beirut using a probabilistic approach *Engineering Science and Technology an International Journal* **23** 373-381

[5] Vono M 2014 Fragility curves of existing RC buildings based on specific structural performance levels *Open Journal of Civil Engineering* 120-134

[6] Zentner I, Gündel M and Bonfils N 2017 Fragility analysis methods: Review of existing approaches and application *Nuclear Engineering and Design* **323** 245-258

[7] Sfahani M.G, Guan H and Loo Y 2015 Seismic reliability and risk assessment of structures based on fragility analysis – A review *Advances in Structural Engineering* **15** 1653-1669

[8] Jeonga S H and Elnashai A S 2007 Probabilistic fragility analysis parameterized by fundamental response quantities *Engineering Structures* **29** 1238-1251

[9] Singh R and Singh V K 2018 Fragility curve for reinforced structural wall *i- Managers Journal on Structural Engineering* **7** 43-49

IOP Conf. Series: Earth and Environmental Science 1149 (2023) 012008 doi:10.1088/1755-1315/1149/1/012008

Mechanical and Microstructure Properties of Plain and Blended Cement Structural Lightweight Aggregate Concrete Using Sintered Fly Ash Aggregates

Pankaj Dhemla[a,b], Prakash Somani[a], Raj Kumar Satankar[c], and B L Swami[a]

[a]Department of Civil Engineering, Malaviya National Institute of Technology, Jaipur, India
[b]Department of Civil Engineering, Poornima Collage of Engineering, Jaipur, India
[c]Department of Mechanical Engineering, Poornima Collage of Engineering, Jaipur, India

Email: pankajdhemla@poornima.org

Abstract. Infrastructure and industry have grown dramatically as a result of globalization and urbanization. However, there are certain challenges, such as meeting infrastructural needs that require a large amount of extraction of natural aggregates and disposal of waste generated as a result of industrialization. Both of these challenges have an adverse impact on the environment. Deforestation, erosion, contamination of nearby streams, and increase in noise, dust, and emissions are all consequences of aggregate extraction. The industrial waste can contaminate groundwater, lakes, streams, rivers and coastal waterways and harm the land and neighboring water bodies. Hence these challenges must be addressed as a priority for environmental sustainability. The current study developed structural lightweight aggregate concrete (SLWAC) using sintered fly ash aggregate (SFA). The investigation was done with the use of different binders such as Portland Pozzolana Cement (PPC), Ordinary Portland Cement (OPC), and Micro Fibre Cement (MFC). The present study explores the suitable binder to produce SLWAC at different water-cement ratios of 0.4, a0.44, and 0.48. The performance evaluation of SLWAC was investigated by mechanical behavior like compressive strength, split tensile, flexural strength, and impact test. Microstructure analysis is also included in this study to validate the mechanical experiment results. SLWAC made from OPC shows more mechanical strength than PPC and MFC, due to faster hydration and early strength gain. SLWAC made from microfiber cement, has more strength as compared to PPC. This could be due to fibers in MFC, acting as a reinforcing agent, which reduces micro cracks. The microstructure of SLWAC shows that the wall effect does not exist in SLWAC compared to normal aggregate concrete. It has a uniform and continuous microstructure of hydration products and partially penetrates the SFA. It could be associated with the superior interfacial transition zone having a lower w/c ratio because of the higher absorption of SFA. The findings show that all developed SLWAC can be used in structural members like beams, columns, and slabs.
Keywords: Sintered fly ash aggregates (SFA); Ordinary Portland Cement (OPC); Portland Pozzolana Cement (PPC); Microfiber Cement (MFC); Interfacial transition zone; Structural lightweight aggregate concrete

1. Introduction

Due to concrete's resilient characteristics, it has become a versatile and second most consumable material after water. However, concrete preparation consumes a large quantity of non-renewable resources, whereas natural aggregates contribute 60 to 70% of the volume of the concrete. Continuous extraction ultimately poses a severe threat to the environment[1]. Therefore, many studies are being performed for the replacement of natural aggregates with more sustainable materials, primarily focusing on the utilization of industrial or agricultural waste to develop artificial lightweight aggregates [2]. Fly ash is an industrial waste generally obtained from coal combustion in thermal power plants. Its disposal and management have become significant problems in recent years. China, the USA, and India are the largest producers of fly ash, where India generates 226 million tons of fly ash annually, out of which 38 million tons of fly ash are unutilized and disposed of in landfills [3]. This vast quantity of unutilized fly ash can be used in the manufacturing of artificial lightweight aggregates, the properties of these lightweight aggregates although depending upon the manufacturing process, type of binders, moisture content, etc. There isa number of methods to produce fly ash aggregates, like cold bonding, autoclaving, and sintering [4]. However, lightweight aggregates made from the sintering process at temperatures 1100 to 1300°c are strong with low water absorption [5][6]. These aggregates are best suited to make SLWAC as density is less than 2000 kg/m^3, and compressive strength is more than 17 MPa as defined by the American concrete institute(ACI) and European Union[7][8][9]. In an endeavour to utilize SFA to its maximum potential, many researchers replaced SFA partially or fully with coarse and fine aggregates observing superior mechanical properties compared to conventional concrete[10][11][12]. Researchers tried to reduce the porosity by decreasing the penetration of water in SLWAC. The addition of supplementary cementitious material (SCM) such as fly ash, silica fume, GGBS, fibre, and stone powder was also opted to decrease the water penetration and enhance the mechanical properties of SLWAC [13][14][15].

Interfacial transition zone (ITZ) has a significant influence on the properties of concrete. In NWC, ITZ has a higher water-cement ratio and inter-particle spacing due to "wall effects" making it weaker, and failure may happen through ITZ [2]. In the case of SLWAC, scanning electron microscope (SEM) examination revealed a nearly continuous and uniform microstructure of hydration products reclining alongside and partially penetrating the SFA [16]. The reason behind superior ITZ could be the higher absorption of SFA, which causes a reduction in the w/cm ratio around the aggregate [17]. Due to the excellent aggregate paste bond in the SLWAC matrix, the fracture plane passes through the aggregate rather than bypassing it, as in NWC. The fracture behaviour of SLWAC is a straight line representing linear elastic fracture mechanics. So special consideration is required for any concentrated load [18]. Thus, SLWAC has advantages like purity, lower dead load, better physical properties, durability, and reduced cost, making it better than conventional concrete [4]. Still, limited use of SLWAC prevails because of lesser manufacturing facilities in India. Standard codes/manuals are not available to check the quality, and it requires careful handling because of the sensitivity of SFA towards moisture [19].

From the above-mentioned literature survey, it can be seen that many researchers tried to develop SLWAC, but the selection of binders in SLWAC and performance with SFA needs attention. This could help in the selection of the right binder with the specific need of the property of SLWAC. The objective of this study was to develop SLWAC with OPC, PPC, and MF binder with water-cement ratios of 0.4, 0.44 and 0.48. The reason behind the selection of these water-cement ratios to develop SLWAC is the desire to achieve a compressive strength between 25 MPa to 40 MPa that can be used in all types of structural members. The fresh, mechanical and microstructure of these mixes were investigated. The flow chart for the proposed research methodology is displayed in Figure 1. Finally, based on these results conclusion is made. This research will promote the use of SFA in construction, which will help dispose of large amounts of unutilized fly ash. It will also save our natural resources, which ultimately help to safeguard our environment. Various types of cement are available in the market, blended with mineral admixtures and fibres. However, there is a limited experimental investigation on blended cement with SLWAC. Therefore, the main objective of the investigation is to know the effects on mechanical properties and durability of SLWAC using ordinary Portland cement

(OPC) and blended cement with different w/c ratios. A comprehensive study was conducted to inspect the mechanical and microstructure characteristics.

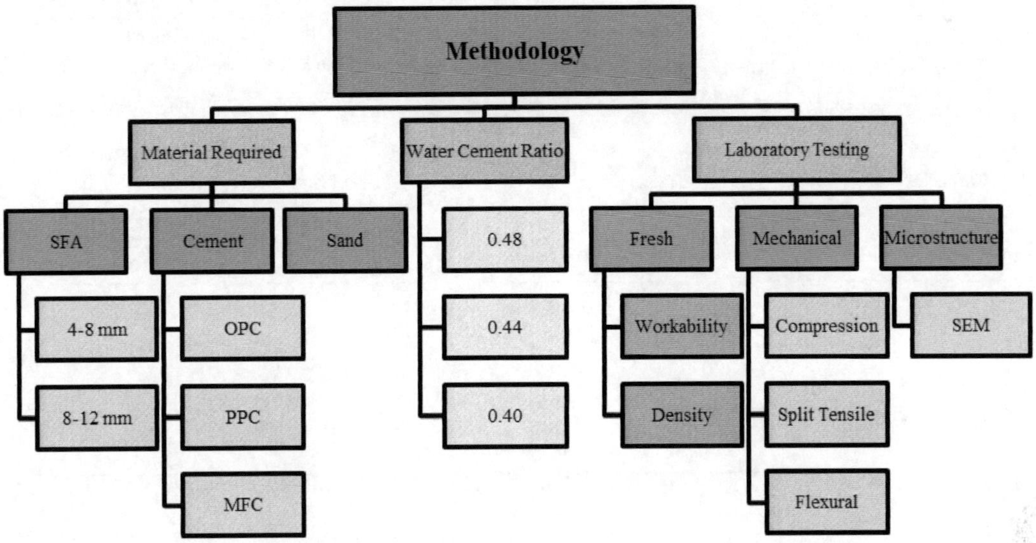

Figure 1. Flow chart for the proposed research

2. Experimental Investigation

2.1. Raw Material

SFA of size 8-12 mm and 4-8 mm were used as coarse aggregates conforming to ASTM C -330 [20]. It mainly comprises 61% Silica, 14% Al_2O_3, 9.0 % of CaO, 8 % Fe_2O_3, and small quantities of alkalis. River sand zone – III is used as fine aggregates. Birla chetak Ordinary Port-land Cement (OPC) - 43 Grade as per Indian Standard IS 8112-1989 [21], Birla Chetak Pozzolana Port-land cement (PPC) and Duraguard micro fiber cement (MF) confirming to IS 1489(part-1) [22] are used as a binder. Sikaplast 5003 is used as a superplasticizer. The physical and chemical characteristics of the binder are tabulated in Table 1 and Table 2. The physical characteristics of SFA are shown in Table 3.

Table 1. Physical properties of binders

Particulars	OPC	IS Code 8112-2013	PPC	MF	IS Code 1489:2015
Fineness m^2/kg	303	**225(min)**	352	333	**300(min)**
Initial setting time (in min)	140	**60**	177	190	**30**
Final Setting Time (in min)	179	**600**	219	250	**60**
Specific Gravity	3.14		2.95	3.10	
Soundness	1.4	**10 (max)**	0.9	0.5	**10 (max)**
Compressive Strength 3 days	35.9	**23**	24.2	24	**16**
Compressive Strength 7 days	42.8	**37.5**	32.8	33	**22**
Compressive Strength 28 days	52.5	**43(min)** **58(max)**	46.5	53	**33**

IOP Conf. Series: Earth and Environmental Science 1149 (2023) 012008 doi:10.1088/1755-1315/1149/1/012008

Table 2. Chemical properties of binders

Particulars	OPC	IS Code 8112-1989	PPC	MF	IS Code 1489:2015
Loss of Ignition	2.00	**5 (max)**	2.63	1.46	**5 (max)**
Insoluble residue	0.30	**2 (max)**	31.94	29.00	**X+4(100-X)/100 where X is declared fly ash**
Mgo (% by mass)	1.50	**6 (max)**	1.80	0.97	**6 (max)**
SO_3 (% by mass)	2.00		2.40	2.17	**3.5(max)**
Declared fly ash			32±3	31.10	**15-35%**

Table 3. Physical properties of the aggregates

Property	Fine Aggregate River Sand Zone - III	SFA Coarse Aggregate (8-12 mm)	(4-8 mm)
Crushing strength (MPa)	-	9.1	9.8
Loose Bulk density (Kg/m³)	-	840	845
Specific Gravity	2.62	1.52	1.53
24 hrs Water Absorption (%)	1.2	16	15
Fineness Modulus	2.52	-	-
Porosity (%)	-	38	39

2.2. Mix Proportioning

The mix design of grade M-25 was adopted from ACI 211.2-98 and evaluated by weight method using a specific gravity pycnometer[23]. The water-cement ratio was kept at 0.48; for the same mix design, different binders at 0.44 and 0.4 water-cement ratios were used. The mixed proportion of this experimental investigation is reported in Table 4. Figure 2 shows cast samples for various tests. SFA were pre-soaked for 30 minutes in 14% of water by weight and mixed in a proportioning of 80% (size 8-12 mm) and 20% (size 4-8 mm). The optimum mix dosages of Sikaplast 5003 were determined by trials in order to achieve a slump of 75± 10 mm.

Figure 2. LWC specimen after casting

Table 4. Mix Proportioning

Mix	Cement (kg)	Natural Sand (kg)	Water (kg)	SFA Aggregate (8-12 mm) (kg)	SFA Aggregate (4-8 mm) (kg)	Plasticizer @0.75% of cement (kg)
OP 40	425	670	170	500	125	3.2
OP 44	386	697	170	500	125	2.9
OP 48	354	727	170	500	125	2.7
PP 40	425	670	170	500	125	3.2
PP 44	386	697	170	500	125	2.9
PP 48	354	727	170	500	125	2.7
MF 40	425	670	170	500	125	3.2
MF 44	386	697	170	500	125	2.9
MF 48	354	727	170	500	125	2.7

2.3. Testing
The experimental program was planned to evaluate fresh properties following the guidelines discussed in IS 1199:1959[24]. For all the mixes used here, samples were cast to inspect the mechanical behaviour, such as compressive, tensile, and flexural, as per the guidelines of IS 516:1959 [25] as well as IS 5816:1999 [26]. The microstructure assessment was performed with a Nova nano field emission scanning electron microscope (FE-SEM-450) which could help to find out the behaviour of the Interfacial transition zone (ITZ).

3. Result & Discussion

3.1. Workability
As discussed in section 2, most of the guidelines illustrate that the ideal SLWAC density lies in the range of 1600-2000 Kg/m^3, and compressive strength of more than 17 MPa is acceptable[7][20]. The study confirms adequate consistency is found in SLWAC made from different cements. The workability of different fresh concrete mixtures was measured with a compaction factor under a fixed dose of superplasticizer. The test outcomes are shown in Table 5, which reveals that blended cement gives higher workability than OPC. This is aligned with the previous research and it is due to presence of more fines in PPC and MF as compared to OPC. Since, finer PPC makes more paste so it becomes more workable as compared to OPC and MF. However, the fiber present in MF cement reduces workability as compare to PPC.

Table 5. Workability of fresh concrete

Types of Binders	**Compaction Factor**		
	w/c ratio (0.40)	**w/c ratio (0.44)**	**w/c ratio (0.48)**
OPC	0.90	0.90	0.91
PPC	0.92	0.94	0.95
MF	0.91	0.92	0.93

3.2. Compressive Strength
The compressive strength of the samples was taken after 7, 14, 28, 56 and 90 days and displayed in Figure 3. It can be observed from the graph that the compressive strength of OPC-40 reached 45 N/mm^2. In all specimens with OPC binder, compressive strength till 14 days was almost 25% more than blended cement. This could be due to more Tri calcium silicate (C_3S) present in OPC, which causes faster hydration and gives early strength. After 28 days, compressive strength with OPC binder

IOP Conf. Series: Earth and Environmental Science 1149 (2023) 012008 doi:10.1088/1755-1315/1149/1/012008

was 10-12 % higher as compared to PPC and MF. It can be suggested that the compressive strength of the samples containing PPC and MF increases faster over a longer duration[27]. The compressive strength of SLWAC containing MF was better than PPC. The reason may be fiber's presence, which increased the density, thereby improving the strength. In their research, Kayali et. al [28] and Zhu et. al [29] obtained similar findings.

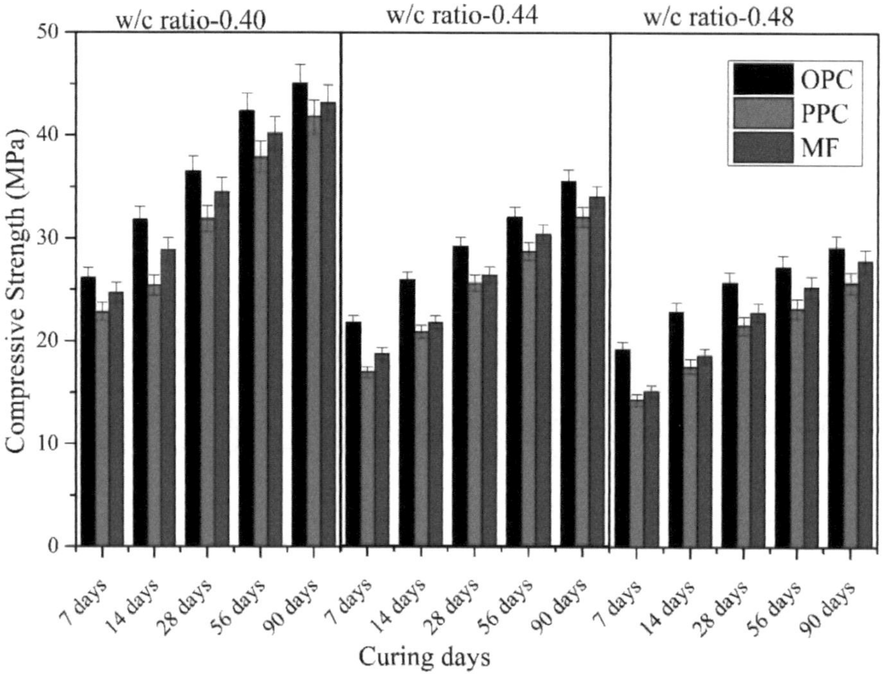

Figure 3. Compressive Strength of LWC specimens at different w/c ratio

3.3. *Split Tensile Strength*

Tensile strength resists shear forces in unreinforced parts, controls cracking in prestressed concrete and resists shrinkage and thermal stress. Figure 4 represents the results of the tensile strength of SLWAC having OPC, PPC and MF as binders at 0.40, 0.44 and 0.48 water-cement ratios, respectively. As per ASTM C330, the minimum tensile strength requirement is 2 MPa for SLWAC and the results of all series of SLWAC with OPC, PPC and MF cement was observed to be greater than 2 MPa, so it can be concluded that the concrete prepared can be used in structural members in construction [20].From the results, it can be inferred that as the water-cement ratio decreases, the tensile strength increases. The split tensile strength of SLWAC of all binders at 0.40 w/c ratio increases by 16-23% from SLWAC of 0.48 w/c ratio at 28 days of curing. Similarly, a 22-28% increment was observed for 56 days of curing.

It is also evident that SLWAC made from OPC gives 3-5% better split tensile strength than MF cement and a 9-12% increment from PPC cement at 28 days of curing. This gap reduces to 2-3% and 7-10%, respectively, at 56 days of curing. The ratio of split tensile strength to compressive strength varies from 0.09 to 0.12, which is lower than NWC [28].

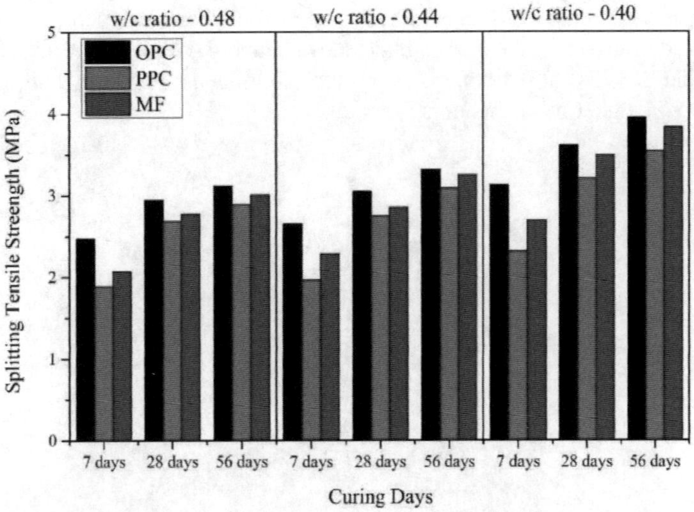

Figure 4. Split Tensile Strength of LWC specimens at different w/c ratio

3.4. *Flexural Resistance*

Flexural strength results of SLWAC made from different binders at 0.4, 0.44 and 0.48 water-cement ratios are depicted in Figure 5. It can be inferred that the flexural strength of all binders 15-20% is more in SLWAC at a water cement ratio of 0.4 than 0.44, and a 23-28% increment is observed from the 0.48 water cement ratio. SLWAC from OPC has flexural strength 2-3 % better than MF, and a 5-8 % increment is observed compared to PPC. The ratio of flexural strength to compressive strength varies from 16-20%, in line with previous research[29].

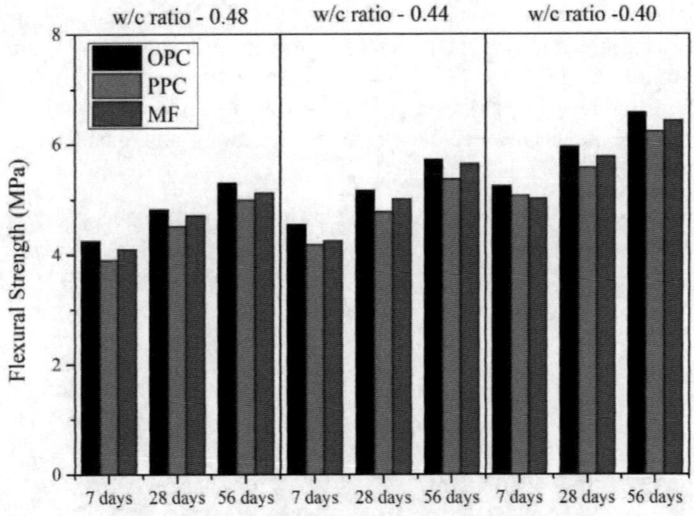

Figure 5. Flexural Strength of LWC specimens at different w/c ratio

3.5. *Bulk Density*

Bulk density also indicates the strength of concrete[30][5].It is evaluated by dividing the weight (in grams) by the volume (cm^3). The graph shows the change in bulk density of SLWAC made from different binders at w/c ratios 0.40, 0.44 and 0.48. It can be inferred from Figure 6 that SLWAC made

IOP Conf. Series: Earth and Environmental Science 1149 (2023) 012008 doi:10.1088/1755-1315/1149/1/012008

from OPC has higher bulk densities than SLWAC made from blended cement. For instance, SLWAC made from OPC binder have 4-4.5% higher bulk density than SLWAC made from MF. Similarly, it can be concluded that SLWAC at 0.4 w/c ratios has more bulk density than at 0.48 w/c ratios. Figure 6 shows it is 3.5 to 4.5 % higher for all binders.

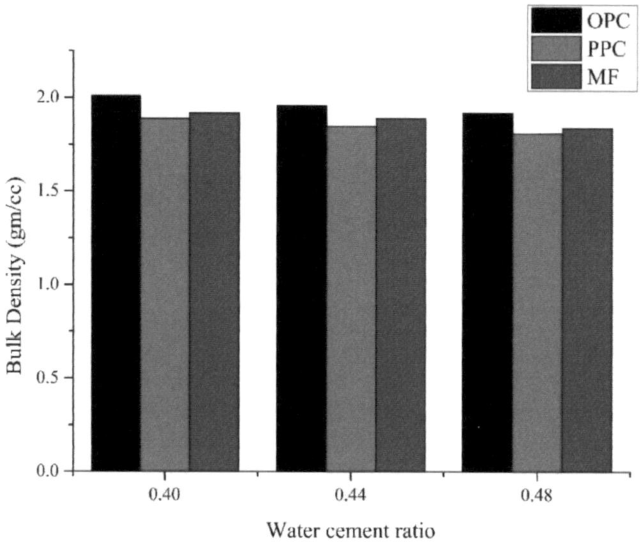

Figure 6. Bulk density of LWC specimen

3.6. Micro Structure Analysis

Specimen size of 10 × 10 × 10 mm was taken out from all 28 days of cured specimens of SLWAC from different binders. These samples were smoothed with abrasive papers of P400, P600, P800, and P1200 and gold coating was done to make them electrically conductive. SEM images were taken by scanning a focused electron beam over the concrete specimen. It is shown in Figure 7 (a), (b), and (c) for SLWAC made from OPC, PPC and MF, respectively, at 0.48 w/c ratio. Similarly, EDX images of SLWAC are shown in Figure 8 (a), (b), and (c) with different binders at a w/c ratio of 0.48. These images help in detecting the occurrence of expansive reactions and quantifying their products in SLWAC.

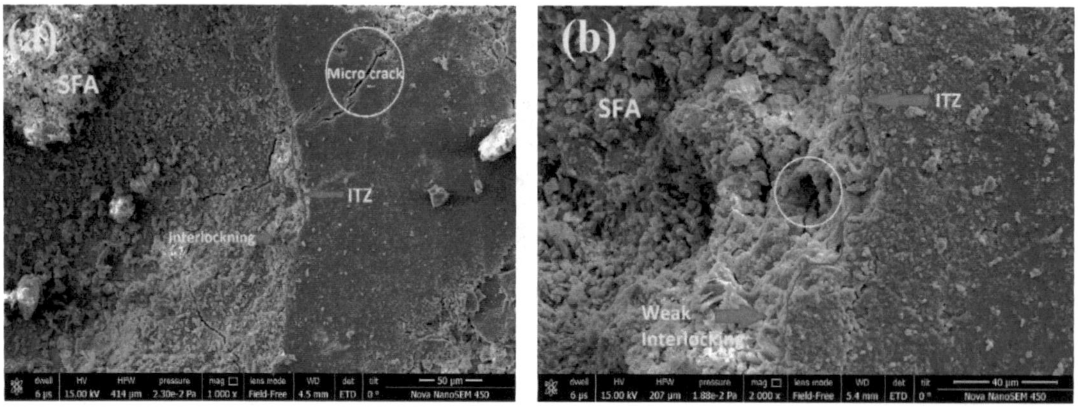

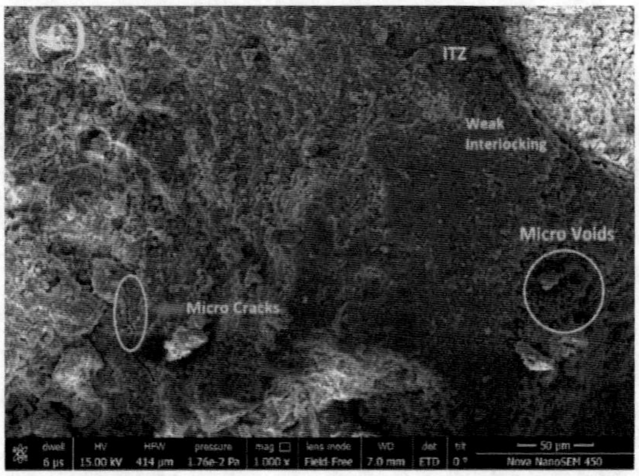

Figure 7. SEM images of LWC made from (a) OPC, (b) PPC and (c) MF

The interaction between different types of cement paste and SFA has been studied to inspect the influence on strength and durability. The interfacial transition zone (ITZ) is the weak link in NWC, but there is no distinct interfacial zone observed in SLWAC that is due to pre-wetted porous aggregates. Cement paste penetrates the porous aggregates, and due to the interlocking between them, the strength of concrete increases[31].

Weak interlocking between aggregate and cement mortar paste was observed in PPC and MF-based cement specimens. This might be due to poor cohesion. Less number of micro voids were detected in OPC-based cement specimens as compared to other mixes [32]. Furthermore, less CSH gel was observed in PPC and MF-based mixes, which might also be a reason for the lower strength of these specimens.

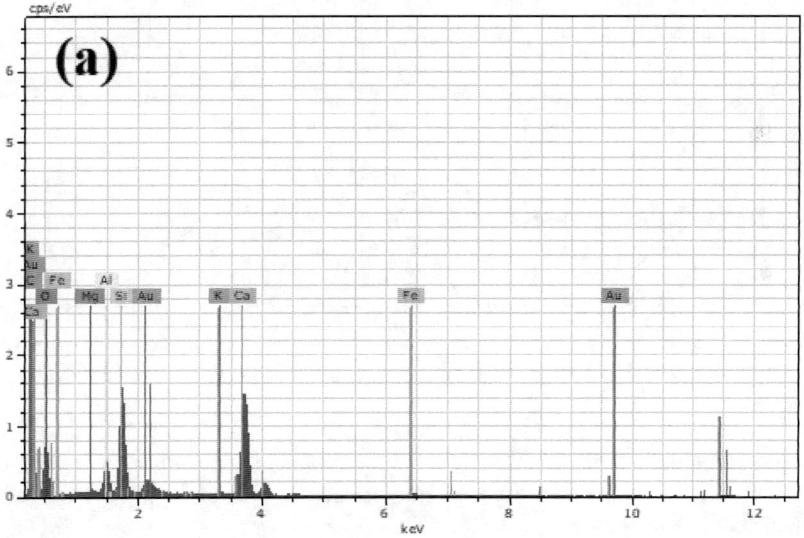

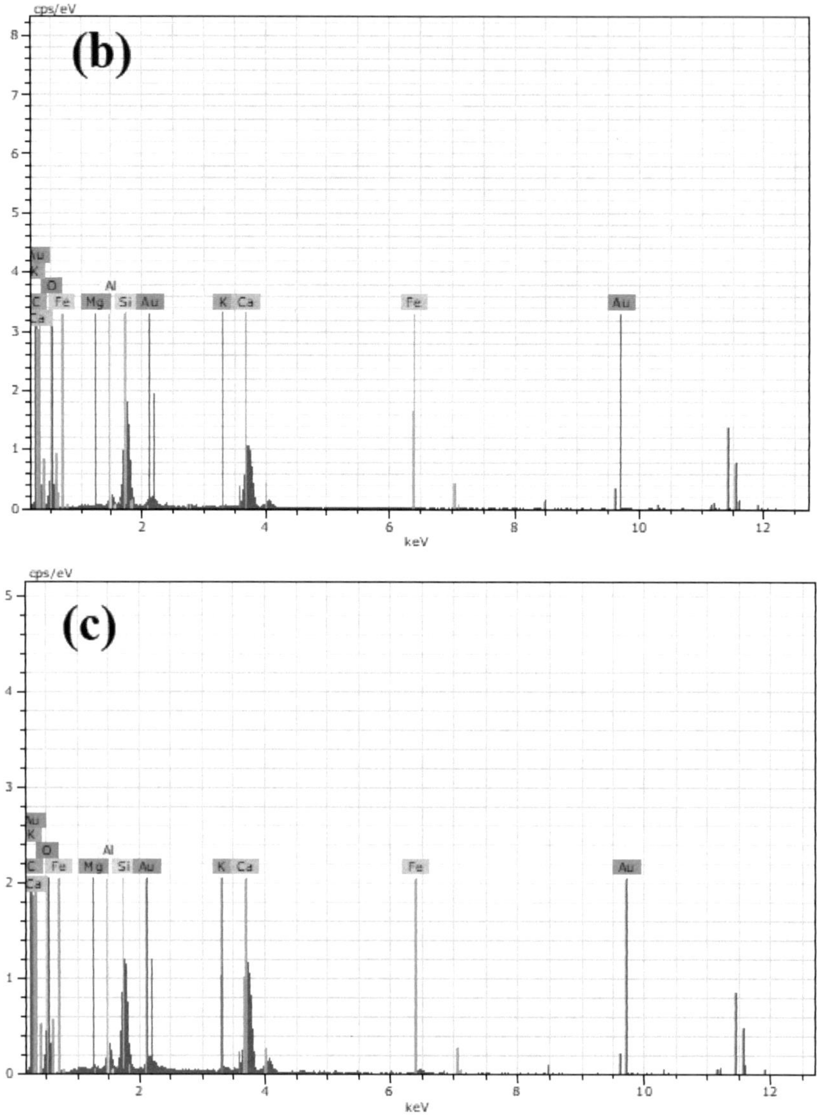

Figure 8. EDX of LWC made from (a) OPC, (b) PPC, and (c) MF

4. Conclusion

After developing three mixes of SLWAC from plain and blended cement, many tests were carried out to evaluate fresh, mechanical, durability and microstructure characteristics. Based on the obtained results following conclusion can be made.

- SLWAC made from OPC shows more mechanical strength than PPC and MF due to faster hydration and early strength gain. On the other hand, SLWAC made from microfiber cement has more strength than PPC; this could be due to the presence of fibres in MF acting as a reinforcing agent, reducing micro-cracks.
- SLWAC absorbs the excess water present in mortar paste, and later on, this water act as an internal reservoir, which causes internal curing, ultimately enhancing the strength.

- The bulk density of SLWAC made from OPC is higher than blended cement. Impact resistance is more of SLWAC made from MF and least with PPC. This is because the fibre along the crack bears more impact than other cement.
- The micro-graph of SLWAC shows that the wall effect does not exist in SLWAC compared to normal aggregate concrete. Instead, it has uniform and continuous microstructure of hydration products and partially penetrates in the SFA; it could be associated with the superior ITZ having less w/c ratio because of higher absorption of SFA.

The strength of SLWAC was good with OPC in 28 days as compared to blended cement. When curing days increases difference of mechanical and durability of SLWAC made with blended cement come closer to made with OPC.

Acknowledgement
The authors would like to thank the Materials Research Center of MNIT Jaipur for their valuable assistance in conducting micro-structural studies.

References
[1] A. M. . J. J. B. Neville, *Concrete technology*. 1987.Pearson Education Limited
[2] M. S. Nadesan and P. Dinakar, "Structural concrete using sintered flyash lightweight aggregate: A review," *Construction and Building Materials*, vol. 154. Elsevier Ltd, pp. 928–944, Nov. 15, 2017, doi: 10.1016/j.conbuildmat.2017.08.005.
[3] A. Yousuf, S. O. Manzoor, M. Youssouf, Z. A. Malik, and K. Sajjad Khawaja, "Fly Ash: Production and Utilization in India-An Overview," *J. Mater. Environ. Sci*, vol. 2020, no. 6, pp. 911–921, 2020.
[4] S. Chandra and L. Berntsson, "Lightweight Aggregate Concrete," 2002. [Online]. Available: www.williamandrew.com.
[5] N. U. Kockal and T. Ozturan, "Effects of lightweight fly ash aggregate properties on the behavior of lightweight concretes," *J. Hazard. Mater.*, vol. 179, no. 1–3, pp. 954–965, Jul. 2010, doi: 10.1016/j.jhazmat.2010.03.098.
[6] P. Gomathi and A. Sivakumar, "Accelerated curing effects on the mechanical performance of cold bonded and sintered fly ash aggregate concrete," *Constr. Build. Mater.*, vol. 77, pp. 276–287, Feb. 2015, doi: 10.1016/j.conbuildmat.2014.12.108.
[7] 213R-2 ACI Committee Report, "Guide for Structural Lightweight-Aggregate Concrete," 2003.
[8] J. M. J. M. Bijen, "Manufacturing processes of artificial lightweight aggregates from fly ash," 1986.
[9] V. Cerny, M. Kocianova, and R. Drochytka, "Possibilities of Lightweight High Strength Concrete Production from Sintered Fly Ash Aggregate," in *Procedia Engineering*, 2017, vol. 195, pp. 9–16, doi: 10.1016/j.proeng.2017.04.517.
[10] M. S. Nadesan and P. Dinakar, "Mix design and properties of fly ash waste lightweight aggregates in structural lightweight concrete," *Case Stud. Constr. Mater.*, vol. 7, pp. 336–347, Dec. 2017, doi: 10.1016/j.cscm.2017.09.005.
[11] H. Z. Cui, T. Y. Lo, S. A. Memon, and W. Xu, "Effect of lightweight aggregates on the mechanical properties and brittleness of lightweight aggregate concrete," *Constr. Build. Mater.*, vol. 35, pp. 149–158, Oct. 2012, doi: 10.1016/j.conbuildmat.2012.02.053.
[12] R. Wasserman and A. Bentur, "PI1 SOOO8-8846(97)00019-7 Effect of Lightweight Fly Ash Aggregate Microstructure on The Strength of Concretes," Elsetier Science Ltd, 1997.
[13] E. Güneyisi, M. Gesoğlu, Ö. Pürsünlü, and K. Mermerdaş, "Durability aspect of concretes composed of cold bonded and sintered fly ash lightweight aggregates," *Compos. Part B Eng.*, vol. 53, pp. 258–266, 2013, doi: 10.1016/j.compositesb.2013.04.070.
[14] N. U. Kockal and T. Ozturan, "Durability of lightweight concretes with lightweight fly ash aggregates," *Constr. Build. Mater.*, vol. 25, no. 3, pp. 1430–1438, Mar. 2011, doi:

10.1016/j.conbuildmat.2010.09.022.
[15] M. N. Haque, H. Al-Khaiat, and O. Kayali, "Strength and durability of lightweight concrete," *Cem. Concr. Compos.*, vol. 26, no. 4, pp. 307–314, May 2004, doi: 10.1016/S0958-9465(02)00141-5.
[16] T. Y. Lo, H. Cui, S. A. Memon, and T. Noguchi, "Manufacturing of sintered lightweight aggregate using high-carbon fly ash and its effect on the mechanical properties and microstructure of concrete," *J. Clean. Prod.*, vol. 112, pp. 753–762, Jan. 2016, doi: 10.1016/j.jclepro.2015.07.001.
[17] L. Kong, L. Hou, and X. Bao, "Application of AC impedance technique in study of lightweight aggregate-paste interface," *Constr. Build. Mater.*, vol. 82, pp. 332–340, May 2015, doi: 10.1016/j.conbuildmat.2015.02.079.
[18] J. Walraven, J. Den Uijl, J. Stroband, N. Al-Zubi, J. Gijsbers, And M. Naaktgeboren, "Structural lightweight concrete: recent research," *Heron (Delft)*, vol. 40, no. 1, pp. 5–30, 1995.
[19] P. Dhemla, B. L. Swami, and P. Somani, "Experimental Investigation of Light Weight Concrete Using Sintered Fly Ash Aggregates," in *IOP Conference Series: Earth and Environmental Science*, 2021, vol. 795, no. 1, doi: 10.1088/1755-1315/795/1/012032.
[20] A. International and files indexed by mero, "Standard Specification for Lightweight Aggregates for Structural Concrete 1." [Online]. Available: www.astm.org,.
[21] Bureau of Indian Standard(BIS), "IS: 8112 – 1989, Specification for 43 grade Ordinary Portland Cement," *Bur. Indian Stand. Delhi*, p. 17, 2013.
[22] B. of Indian Standards, "IS 1489-1 (1991): Specification for Portland pozzolana cement, Part 1: Flyash based."
[23] J. R. Prestera Secretary Michael Boyle Chairman David A Crocker Chairman *et al.*, "ACI 211.2-98 supersedes ACI 211.2-91 and became effective," 1998.
[24] I. 1199 (1959), "Methods of sampling and analysis of concrete."
[25] B. of Indian Standards, "IS 516 (1959): Method of Tests for Strength of Concrete."
[26] B. of Indian Standards, "IS 5816 (1999): Method of Test Splitting Tensile Strength of Concrete."
[27] A. K. Parande, B. R. Babu, K. Pandi, M. S. Karthikeyan, and N. Palaniswamy, "Environmental effects on concrete using Ordinary and Pozzolana Portland cement," *Constr. Build. Mater.*, vol. 25, no. 1, pp. 288–297, Jan. 2011, doi: 10.1016/j.conbuildmat.2010.06.027.
[28] V. Saraswathy, S. Karthick, H. S. Lee, S. J. Kwon, and H. M. Yang, "Comparative Study of Strength and Corrosion Resistant Properties of Plain and Blended Cement Concrete Types," *Adv. Mater. Sci. Eng.*, vol. 2017, 2017, doi: 10.1155/2017/9454982.
[29] J. N. Farahani, P. Shafigh, B. Alsubari, S. Shahnazar, and H. Bin Mahmud, "Engineering properties of lightweight aggregate concrete containing binary and ternary blended cement," *J. Clean. Prod.*, vol. 149, pp. 976–988, Apr. 2017, doi: 10.1016/j.jclepro.2017.02.077.
[30] H. Al-Khaiat and M. N. Haque, "Effect Of Initial Curing On Early Strength And Physical Properties Of A Lightweight Concrete," 1998.
[31] M. S. Nadesan and P. Dinakar, "Micro-structural behavior of interfacial transition zone of the porous sintered fly ash aggregate," *J. Build. Eng.*, vol. 16, pp. 31–38, Mar. 2018, doi: 10.1016/j.jobe.2017.12.007.
[32] M. S. Nadesan and P. Dinakar, "Influence of type of binder on high-performance sintered fly ash lightweight aggregate concrete," *Constr. Build. Mater.*, vol. 176, pp. 665–675, Jul. 2018, doi: 10.1016/j.conbuildmat.2018.05.057.

Rubberized Geopolymer Mortar and Concrete: A Comprehensive Review.

Arjun Raj P K[1], Sarath D[2], Dr. Praveen Nagarajan[3] and Dr.A. P. Shashikala[4]

[1]Research Scholar, Department of Civil Engineering, National Institute of Technology, Calicut, India
[2]M tech Scholar, Department of Civil Engineering, National Institute of Technology, Calicut, India
[3]Professor, Department of Civil Engineering, National Institute of Technology, Calicut, India
[4]Professor, Department of Civil Engineering, National Institute of Technology, Calicut, India

Corresponding author E-mail: arjun30746@gmail.com

Abstract. In many crucial structural applications, a material's ability to withstand impact is of the utmost importance. The use of crumb rubber particles as a partial replacement of natural aggregates forms rubberized geopolymer concrete (RuGPC), which is found to have great impact resistance and energy absorption capacity. Geopolymer paste and natural aggregates alone are weak in handling the impact loads. The field of Civil engineering has been looking into ways for the last few years to use solid waste and ecologically friendly raw materials as components of concrete in building to embrace sustainability. Greenhouse gas emissions from cement manufacturing and the depletion of natural aggregates are two major issues the construction industry is currently experiencing. RuGPC, which combines the benefits of geopolymer concrete (GPC) and rubberized concrete to create a practical, sustainable building material, has been a hot topic for the past few years. In this paper, the fresh properties, strength and durability characteristics, mixing procedure and curing properties, dynamic and impact properties, microstructures and thermal properties of rubberized geopolymer mortar and concrete are comprehensively reviewed.

Keywords: Rubberized geopolymer Concrete, waste rubber tyres, strength and durability properties.

1. Introduction

Waste rubber tyre disposal has grown to be a serious issue on a global scale. Tires are frequently buried or disposed of, posing a major hazard to the environment [1, 2]. Globally, it is anticipated that more than 1.5 billion end-of-life tyres (ELTs) comprising more than 40% of vulcanised rubber will be disposed of each year [3]. The amount of waste produced globally is predicted to increase by 70% to 3.5 billion tonnes during the next thirty years [4]. Currently, large amounts of tires—3 billion in the EU and 1 billion in the US—are either already stacked (tyres) or landfilled (shredded tyres). By 2030, it is estimated that there will be 1200 million more tyres from cars or 5000 million more tyres that will be regularly disposed of. The disposal of used tyres poses a severe ecological risk. The main sites where used tyres are disposed of are responsible for the decline in biodiversity. The tyres also contain hazardous and soluble materials [5]. There are many ways by which tyres can be disposed of including landfill disposal, burning, use as fuel pyrolysis etc. causing environmental issues. One possible method of recycling these waste tyres is to use them as aggregates in concrete as well as geopolymer concrete. By adopting this method, problems caused due tyre burning or landfill of tyres and depletion of natural resources can be eliminated.

Portland cement concrete has become an essential requisite for the construction industry. After water, it is the material that people utilize the most [6] and its annual production is predicted to be more than 10 billion tonnes [7]. Generation of Portland cement is a significantly energy-consuming operation that uses a large amount of energy, about 2% to 3% of the world's primary energy [8]. In addition, it is estimated that for each tonne of cement delivered, about one tonne of CO_2 is produced, accounting for 7% of the CO_2 emitted into the environment. [7]. To tackle this situation of CO2 emission and high energy requirement in the production process, alternative binders can be used. Geopolymer concrete often referred to as alkali-activated concrete is an efficient approach to reform conventional concrete environmentally friendly by utilising less or no Portland cement [9]. According to reports, using geopolymer production in place of cement would reduce CO2 emissions by as much as 80% [10]. A novel class of inorganic polymer called alkali-activated geopolymers is made up of [SiO4] and [AlO4] tetrahedrons connected by atoms of oxygen [11]. Due to its superior qualities and minimal influence on the environment, geopolymer has been considered a viable substitute for OPC. The strength in compression, durability, strength in interface bonding of geopolymer have all been reported to be higher than those of OPC in prior studies. Compared to conventional concrete, GPC has a considerably lower carbon footprint. Greenhouse gas emissions can be cut by 50% if cement is fully replaced with a geopolymer [12, 13]. Alumino-silicate base materials, which are utilised to produce geopolymer, are widely accessible, which is another advantage of using geopolymer. This means that waste materials including fly ash, GGBS [14], Silica fume [15], Waste wood ash [16], metakaolin [17], Rice husk ash[18] etc. can be utilized to prepare geopolymers, which is a viable method of waste recycling.

To address the problems brought on by CO2 emissions and the depletion of natural aggregates, rubberized geopolymer concrete (RuGPC), which combines the advantages of rubcrete and geopolymer concrete, can be employed. One of the main disadvantages of using rubber in geopolymer concrete is the reduction of mechanical properties [14] but the durability in terms of resistance to abrasion [14] and impact strength of RuGPC is reported to be superior to the reference specimens [19]. By treating rubber particles with chemicals like NaOH, potassium permanganate, etc., the problem of deterioration in mechanical properties can be partially avoided [20, 21]. NaOH is already using as an alkali activator for alumino silicate base materials for geopolymer. In this work, the fresh properties, strength and durability characteristics, fabrication and curing properties, dynamic and impact characteristics, microstructures and thermal characteristics of rubberized GPC and mortar are comprehensively reviewed. The work also emphasises environmental impact, classifications and properties of waste rubber and cost analysis of RuGPC. Applications of RuGPC, research gaps, and suggestions for future investigations are all critically examined.

2. Constituents of RuGPC

Fly ash, GGBS, Silica fume, Waste wood ash, metakaolin, Rice husk ash, Red mud etc are the alumino-silicate base materials used for the preparation of Rubberized geopolymer concrete. These waste materials release silica and aluminum when they are combined with alkaline activators like NaOH and Na_2SiO_3, which results in the creation of geopolymer. The nature and reactivity of geopolymer formed will depend upon the chemical constituents of these alumino-silicate base materials [22]. Table 1 shows the chemical composition of aluminosilicate base materials employed in order to create rubberized geopolymer mortar and concrete. Table 2 depicts alumino-silicate base materials, curing methods, admixtures and alkaline solutions used for the preparation of RuGPC concrete and mortar.

From tables 1 and 2, it is revealed that most studies use fly ash, GGBS or a combination of both of them. In the case of fly ash, class F and class C are used. But class F is used more because of the favourable setting properties of fly ash.

In the case of alkaline activators, a combination of NaOH and Na_2SiO_3 is mostly used for the preparation of RuGPC [28, 23, 27, 30, 29, 19, and 38]. From table 2, the ratio of alumino-silicate base materials to alkali ranges from about 0.3 to 0.85 with an optimum value range of 0.4-0.6. The molarity of sodium hydroxide used for the study ranges from 10M to 20M with an optimum value range of 10M-14M. The flow value of geopolymer mortar with crumb rubber was found to increase in proportion to the alkali to alumino-silicate base materials ratio [30]. Also, the strength in compression of crumb rubber geopolymer mortar decreased with respect to alkali to binder ratio according to Wongsa et al. [30].

Table 1. Chemical composition in % of aluminosilicate base materials

Chemical composition %														Ref.
Fly ash														
SiO_2	Al_2O_3	Fe_2O_3	MnO	MgO	Na_2O	SO_3	CaO	K_2O	P_2O_5	TiO_2	CL	SrO	LOI	
55.9	23.9	7.9	0.1	1.3	0.4	0.3	7	1	0.5	1.3	-	-	0.3	[14]
24.8	18.7	2.1	-	0.70	-	-	0.73	0.7	-	1.0	-	-	-	[16]
39.7	11.15	1.20	-	7.61	-	0.46	34.38	-	-	-	-	-	-	[19]
62.55	29.02	4.22	-	-	0.19	0.22	3.12	-	-	-	-	-	0.52	[20]
39.4	20.8	11.5	-	2.2	1.4	4.2	14.5	2.4	0.2	0.5	-	-	1.5	[30]
50.30	22.90	8.17	0.08	2	-	0.58	3.38	3.55	-	1.15	-	-	-	[32]
55.9	27.8	7.09	-	-	-	-	3.95	1.55	-	2.25	-	0.37	-	[39]
GGBS														
31.4	13.1	0.8	0.2	5.5	0.3	4	43.2	0.3	-	0.6	-	-	0.6	[14]
39.79	11.15	1.20	-	7.610	-	0.46	34.38	-	-	-	0.01	-	-	[19]
32.92	13.8	0.58	-	5.76	0.2	3.33	42.13	0.32	0.03	0.57	0.01	-	0.2	[23]
35.12	14.20	0.62	0.69	8.47	0.98	-	39.08	-	-	0.71	-	-	0.13	[31]
34.10	12.30	0.41	0.25	8.12	-	2.59	44.20	0.56	-	0.96	-	-	-	[32]
36.1	9.3	0.03	-	8.9	0.8	2.2	39	0.6	-	-	-	-	1.01	[36]
36.95	10.01	1.48	0.52	6.43	1.39	3.52	33.07	0.74	0.1	0.52	0.05	-	-	[37]

The aggregates used for the study include fine and coarse natural aggregates classified based on size. Table 2 shows that except Azmi et al.[40], most research used river sand as fine aggregates. RuGPC is made using a range of coarse aggregates, including crushed basalt [28], crushed dolomite [19], and crushed gravel [17, 38]. The aggregate fraction in rubberized geopolymer mortar is entirely made up of fine aggregates. Copper slag was utilised by Ameri et al. [36] as a fine aggregate and replaced 5%, 10%, and 15% of the volume with crumb rubber from the same zone as the copper slag.
Rubber aggregates of various sizes are used to replace the fine, coarse, or both fractions. Coarse rubber aggregates include rubber fibres of suitable size, shredded rubber particles [19, 23, 38] etc. Crumb rubber particles are frequently used to replace fine aggregate proportion [23, 27, 30, 19, and 39]. Mechanical properties are found to decrease when the aggregate proportion is replaced by rubber particles [14]. Different types of treatment techniques are introduced for roughening the rubber's surface, enhancing the bond between the geopolymer matrix and the crumb rubber and overcoming strength decrement [20, 21]. Lazorenko et al.[21] treated the crumb rubber particles with NaOH solution, 10% H_2SO_4 solution, UV radiation, 99% pure acetone solution and 5% $KMnO_4$ solution. They reported that $KMnO_4$ treated rubberized geopolymer concrete showed a considerable improvement in compressive strength in comparison with rubberized geopolymer with no treatment.

Generally, two types of admixtures are used in rubberized geopolymer concrete. One is polycarboxylate based [32, 23] and the other one is naphthalene based [28, 29, 38]. According to Montes et al. [44], the addition of superplasticizer based on polycarboxylate marginally decreased the paste's viscosity whereas the addition of superplasticizer based on naphthalene sulfonate slightly increased the paste's viscosity and caused flash set. Superplasticizer dosages used to prepare RuGPC range from 1% [32] to 2% [29]. The strength in compression of rubberized concrete was studied by Luhar et al. [45] who found that higher concentrations of naphthalene sulfonate-based superplasticizer resulted in lower compressive strength.

3. Curing and mixing procedures for RuGPC

Different curing methods adopted for RuGPC are steam curing, ambient curing, water curing and oven curing. Geopolymer mixes will become chemically unstable in the presence of water and geopolymer concrete is not commonly cured using water because alkalies in the specimen will be leached out [46]. Samples are heated to temperatures between 60°C and 90°C for 24-72 hours during the heat-curing procedure, after which they are returned to room temperature until testing [28]. Steam curing includes heating samples to a temperature of about 46°C for seven days in an autoclave. After that, the samples are removed and left to cure at ambient temperature until the test [25]. Oven-cured crumb rubber geopolymer mortar specimens at 60°C and 90°C showed superior compressive and flexural strength in comparison with ambient cured specimens [30].

Table 2 Alumino-silicate base materials, curing methods, alkalies, aggregates and admixtures.

Ref	Aluminosilicate base materials	Alkalies	Alkali to binder ratio	Type of Admixture and dosage.	Type of curing and duration	Aggregates	Rubber particles
[28]	Class F fly ash	NaOH-flakes-14M Na_2SO_4-n r Na_2SO_4/NaOH-2.5	0.4	Naphthalene sulphonate based-2%	Oven curing-90°C-48h	FA-River sand (35%), sp.g-2.61. CA-crushed basalt-10,20mm sp.g-2.59	Fibres- 2-4 mm width, sp.g-1.09. Replacement - 10% by weight of FA.
[23]	Class F fly ash GGBS	NaOH-pellets-14 M Na_2SO_4- n r Na_2SO_4/NaOH-2.5	0.4-0.6	Polycarboxylate-based and VMA	Ambient-23°C-7-28days	FA-Silica sand(45%) less than 4mm. CA-crushed aggregates-7-10 mm size	Crumb rubber-2-5 mm, 5-10mm, sp.g-1.15, Replacement -10, 20% of FA and 20% of CA. Water soaking treated.
[30]	Class C fly ash	NaOH-10M,15M, 20M Na_2SO_4-2.41 M.S Na_2SO_4/NaOH-	0.65, 0.75,0.85	nr	Oven-60°C 90°C- 48h	Water absorption 0.35 percent, bulk density 1671 kg/m3, and FA-River sand-Sp gravity 2.63 .	For crumb rubber, these parameters are: sp gravity: 1.16, bulk density: 487 kg/m3, water absorption: 3.27%, and FM: 3
[29]	Class F fly ash	NaOH-14M Na_2SO_4- nr Na_2SO_4/NaOH-2.5	0.4	Naphthalene sulphonate based-2%	Oven curing-90°C-48h	Water absorption 0.5%, Sp gravity 2.61 FM is 2.56, CA- 10mm, 20mm	Sp.g: 1.09, 22 mm in length, 2-4 mm in width, and 8–10 aspect ratio for the fibres. 10%, 20%, and 30% FA replacement by weight
. [19]	GGBS	NaOH-nr Na_2SO_4-3.39 M S Na_2SO_4/NaOH-0.39	0.54	nr	Room Temperature-until testing	FA-Natural, pure (35%) sand, less than 0.5 mm in size. 2.65g/cc Sp. FM: 2.25. CA-Dolomite, clean crushed, size 12 mm, specific gravity 2.96 g/cc.	Crumb rubber of mesh 40 sizes and 1-4 mm size. Sp g: 0.45. Replacement : 10%, 20%, and 30% by volume of both FA and CA.

[38]	Class F fly ash	NaOH-10 M Na_2SO_4-nr	0.45	Naphthalene sulphonate based	Curing in Hot water-60°C	2 and 4 mm FA-River sand CA: Maximum size 10 mm crushed gravel.	Recycled rubber pieces in sizes 2 and 4. 10-20-30% volume replacement with two sizes.
[39]	Class F fly ash	NaOH-12 M Na_2SO_4-nr Na_2SO_4/NaOH-2.5	0.5	nr	Curing in Seawater-until testing	FA: River sand: size less than 4.75 mm. CA: Crushed stone: Max size of 20 mm.	Rubber crumbs with sizes 0.5 cm and 1cm. Replacement : 5%, 10%, and 15% of rubber crumbs by weight.
. [37]	GGBS	NaOH-nr Na_2SO_4-3.39 MS Na_2SO_4/NaOH-nr	0.3	nr	Oven Curing 45°C-28 days	nr	Rubber powder: sp g:1, Replacement : 1%, 2.5%, 5%, 10%, 15%, of slag by weight.
. [24]	Class C fly Ash	NaOH-10 M Na_2SO_4-2.41 MS Na_2SO_4/NaOH-1	0.75	nr	Oven-60°C-48hr	River sand: Bulk density: 1670 kg/m3, FM: 3.30, water absorption: 0.35% and sp. g: 2.63.	Crumb rubber: 487 kg/m3, FM:3, water absorption: 3.27%, Sp.g: 1.16 Replacement : 25%, 50%, 75% and 100% by volume of river sand.
. [32]	Class F fly ash GGBS	NaOH-10M Na_2SO_4-2 MS Na_2SO_4/NaOH-2	0.4	Polycarboxylate based-1%	Ambient-until testing	FA: River sand:	Crumb rubber: Size: 2-6 mm, 1.68 mm, 0.595 mm. Replacement 2-6 mm-10% by weight of sand. 1.68 mm-50% by weight of sand 0.595 mm-40% by weight of sand.
. [17]	Metakaolin Kaolin	NaOH-15M Na_2SO_4-nr Na_2SO_4/NaOH-2.9	0.84	nr	Oven-65°C-48hr	FA: Natural sand CA: Crushed gravel: 4.75mm-12.5 mm	Waste rubber powder: Size less than 1 mm Replacement : 5%, 10%, 15% by

							volume of FA.
[40]	Fly ash GGBS	NaOH-12 M Na_2SO_4-nr Na_2SO_4/NaOH-nr	nr	nr	Ambient-until testing	FA: Dune sand with d50 of 240 µm CA: 4mm ungraded and 7mm graded aggregates	Crumb rubber: 2-5mm and 5-7 mm. Replacement : 15% and 30% by weight of CA. Treatments: Water treatment, NaOH treatment, Activator treatment.
. [14]	Class F fly ash GGBS	NaOH-12 M Na_2SO_4-2.3 MS Na_2SO_4/NaOH-1.58	0.37	nr	Ambient - until testing	nr	Replacement : 5%, 15%, 20% replacement of FA.
[40]	Class F fly ash	NaOH-10 M Na_2SO_4-nr Na_2SO_4/NaOH-2	0.4	nr	Room Temperature until testing	FA: Natural sand Zone 2	75micron to 4.75 mm. 5% sand by weight. Hydroxide treatment is used

Wongsa et al.[30] reported that the early age strength in compression of the specimens was substantially raised by heat curing at 60°C and 90°C for 48 h; as a result, the strength development rate with ageing was slower for the heat-cured specimens than for the ambient cured specimens. The compressive strength of RuGPC samples was examined by Luhar et al. [45], who found that 48 hours of curing was the ideal amount of time. Because ambient curing uses less energy, shrinks concrete less, and produces fewer surface cracks than heat curing, it is a desirable alternative to heat curing.

The alkaline activators used for the preparation are made one day before the casting procedure [38]. For the first two to five minutes, dry components including crumb rubber particles, aggregates, and aluminium silicate base materials were mixed. Alkalies and additional water have been added to the dry components and blended for 2–5 minutes. Following that, admixtures were gradually added and blended for a further one to two minutes. Concrete was then poured into the prepared moulds, and samples were frequently compacted into layers of two to three using a vibrating table to generate dense samples. Aluminosilicate precursor and alkalies were blended for 3 to 5 minutes in studies on RuGPC, followed by the addition of aggregates and rubber particles, mixing for an additional 5 minutes, casting and compaction[39].

4. Fresh properties and physical properties

4.1 Setting time, Workability, Density and Water absorption

The general trend is for the mix's workability to deteriorate as the percentage of rubber particles in concrete or mortar samples increases [31, 36, 30, 33]. Wongsa et al. [30] reported a decrease of 45.4% in flow value when 100% of sand is replaced by crumb rubber. Similarly, Ameri et al. [36] observed a decrease of 14%, 29%, and 50% in flow diameter when fine aggregates are replaced by crumb rubber. Long et al.[32] tried some higher replacement percentages of 20, 40 and 60 and obtained a reduction of flow diameters of 8.3%, 22.22% and 33.33% respectively. The increased flow resistance of rubberized mixes over control mixes may be caused by the crumb rubber's surface roughness. The

contaminants on the crumb rubber aggregate's surface could also be a contributing factor in the decreased workability [36].
With the addition of rubber particles, GPC's setting times increase. Arunkumar et al. [16] showed that RuGPC mixes can have setting times that are increased by 36% and 22%, respectively, by adding 2% rubber fibres.
With an increase in the replacement of rubber particles, the density of rubberized geopolymer concrete generally declines [34]. Table 3 shows the effect of crumb rubber replacement on the density, porosity and water absorption of concrete and mortar specimens.

Table 3 Crumb rubber replacement effect on the density, porosity and water absorption of RuGPC specimens.

Ref	Rubber aggregate size (mm)	Level of Replacement	Properties of specimen	Effect on density	Effect on porosity and water absorption	Comments
[39]	5 to 10 mm	5,10, 15, 20% of FA	Cubic ($Na_2SiO_3/NaOH$=2-2.5,Molarity of NaOH= 12M)	2.96%-7.2% Reduction	5.88%-123% increase in water absorption	The largest weight change increment (1.9%) is seen at 20% replacement, which is also a result of incorrect compaction during moulding.
[15]	1-3 mm and 5-7 mm.	0, 15% and 30%	Cylindrical ($Na_2SiO_3/NaOH$=2.5, Molarity of NaOH= 12M)	9.18%-15.38% Reduction	nr	Crumb rubber's low specific gravity led to a drop in density.
[40]	nr	5,10, 15, 20% of FA	Cubic Mortar ($Na_2SiO_3/NaOH$=2,Molarity of NaOH= 12M)	12.12%-13.93% reduction	3%-12% increase in water absorption	When the water absorption is greater, the mortar's density is lower. This is a result of the concrete having more voids than expected.
[30]	Max 4 mm	0, 100% of FA	Cubic Mortar ($Na_2SiO_3/NaOH$=2,Molarity of NaOH= 12M)	42% Reduction	Water absorption is 5.7 times greater than in the control specimen, and porosity is 1.5 times higher.	The differing porosities of crumb rubber and sand in this investigation may have an impact on the mortar's porosity. The porosity of the rubberized mixes rose as a result of adding rubber particles to mortar, which also increased the amount of air trapped with in mixture.

5. Mechanical Properties

5.1 Compressive strength

The compressive strength gradually decreased as the amount of crumb rubber rose [5, 14, 20, 26, 29, and 42]. Figure 1 depicts the scatter plot of normalised compressive strength changes with various amounts of fine aggregate replacement with crumb rubber. Rubber aggregates and geopolymer paste produce an interfacial transition zone that is weak as a result of their weak bonding, which is the main factor reducing compressive strength [15, 42]. Additionally, the paste of cement with rubber crumbs would wrap the aggregate and be significantly softer than it would be without rubber. The specimens fail quite soon resulting from the cracks quickly developing around the rubber particles while loading

[5]. When rubber particles and geopolymer paste are not properly bonded, applied forces are not distributed evenly, which can cause cracks [5, 20, 40]. Another reason for lower compressive strength is rubber's susceptibility to rising during vibration as a result of its low specific gravity and deficiency of adhesion to other concrete components, which results in the top layer having a larger concentration of rubber ending up in a non-homogeneous concrete[5,42].
Under compression, there was no brittle failure of the rubberized concrete, in contrast to the control specimens. For the rubberised concrete specimens, horizontal cracks were seen, whereas, for the control specimens, inclined cracks were seen [5]. It is also seen that the pre-treated crumb rubber geopolymer concrete exhibits improved performance [20]. According to previous research, the normalized compressive strength (ρ) of RuGPC at various rubber replacements ϕ can be found by the following equation.

$$\rho = e^{-0.026\phi} \quad (1)$$

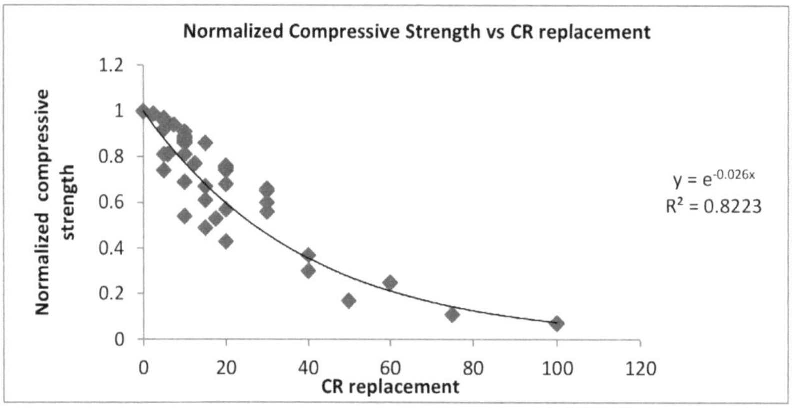

Figure 1 Scatter plot of variation in compressive strength with various percentages of fine aggregate replacement with crumb rubber [5,14,15,17,30-39,42]

5.2 Split tensile strength

The reference mix possessed the greatest split tensile strength and strength steadily decreased as the percentage of rubber increased. [19, 20]. Figure 2 depicts the variation in split tensile strength with various percentages of fine aggregate replacement with crumb rubber. As per Aly et al. [19] a reduction of 34.60%, 23%, and 35.50% of splitting strength was observed corresponding to 10%, 20%, and 30% crumb rubber replacement. But for Saloni et al. [20] and Toufigh et al. [42] a gradual strength decline is observed for the corresponding increase in crumb rubber proportion.
The reason behind the dropping of splitting tensile strength could be the same as the reason for the decreased compressive strength, as well as the small rubber particles play in separating the aggregates from one another and also the binder paste, which results in weak interparticle adhesion in the mix [19]. However, the pre-treatment of crumb rubber can improve the splitting tensile strength [20].

5.3 Flexural Strength

The flexural strength gradually declined as the percentage of crumb rubber increased [5, 19, 20]. The change in flexural strength is shown in Figure 3 with various percentages of aggregate replacement with crumb rubber. Aly et al. [19] as well as Saloni et al. [20] shows a gradual reduction in flexural strength with varying percentage of crumb rubber.

Additionally, it was noted that rubcrete didn't display brittle failure under flexural tensile loading while the control specimens did and were broken into two pieces. The rubber containing specimens demonstrated deformation without completely collapsing, in contrast to the control specimens, which exhibited brittle failure and split into two pieces right after cracking. [5].

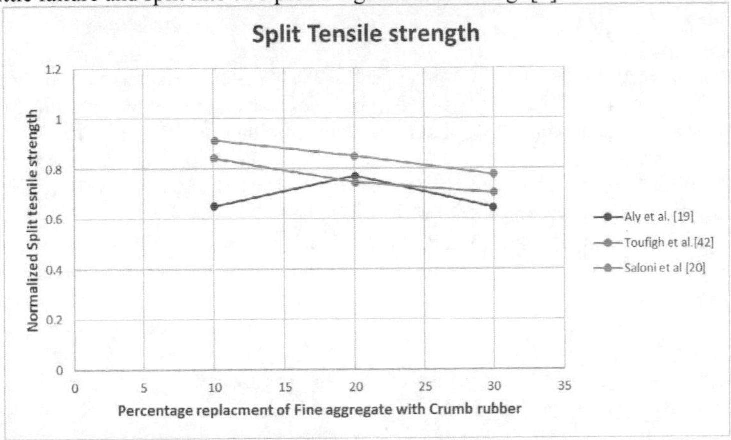

Figure 2 Scatter plot of variation in split tensile strength with various percentages of fine aggregate replacement with crumb rubber

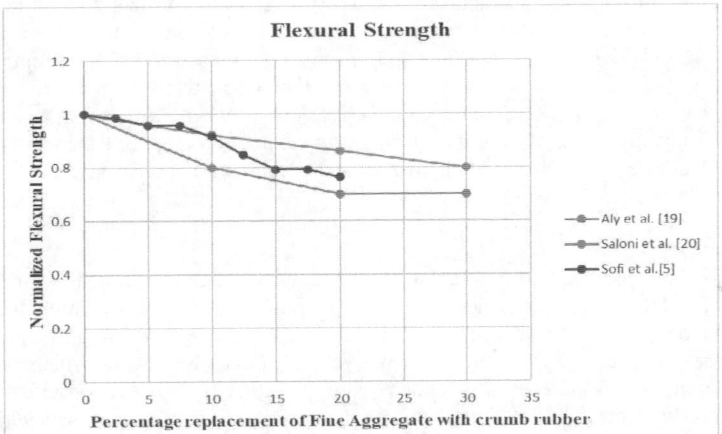

Figure 3 Scatter plot of variation in Flexural strength with various percentages of fine aggregate replacement with crumb rubber

5.4 Impact resistance properties

Following ACI-544, Aly et al. [19] conducted a drop weight test utilising cylindrical RuGPC discs (150x65 mm). A steel ball of weight 4.5 kg was required to be released from a depth of 450 mm to start the test. The number of impacts required for the first crack and failure was measured. The study showed that more strikes are required to form initial and final cracks as the amount of crumb rubber increases, implying that rubberized specimens have high impact resistance. The number of impacts for the initial visible crack and the number of blows for ultimate failure increased by 3.0 and 2.50 times, respectively, as the percentage of crumbs increased from 0 to 30% in geopolymer concrete mixtures [19]. Similar experiments were conducted by Erdem et al. [48] using a steel ball with a weight of 5 kg that was dropped from a height of 1 m. For GPC mixtures comprising, respectively, 0%, 20%, and 50% rubber percentage, the first cracks appeared at 3, 5, and 7 strikes. The final hits were 4, 17, and 38 with collapse impact energies of 250 kNmm, 900 kNmm, and 1900 kNmm for each mix. When exposed to impact loading, rubber particles can enhance impact strength and help to avoid fracture formation, in contrast to geopolymer paste and natural aggregate. [48, 43].

6. Durability Properties

According to Luhar et al. [29], RuGPC samples had better abrasion resistance in terms of wear depth after 28 days when more crumb rubber was used in place of fine aggregate. For instance, after 28 days, the wear depth was 1.28 mm with 0% rubber fibre substitution and 0.8 mm with 30% rubber fibre substitution, respectively. An abrasive powder's influence on the surface is weakened because rubber particles lengthen more than the smooth surface and work as brushes when abrasive forces are acted, contributing to the partial increase in abrasion resistance [29]. Saloni et al.[20] investigated the resistance of RuGPC specimens against acids such as HCl and H_2SO_4. They showed that for crumb rubber replacement of 10%, 20% and 30%, the compressive strength loss percentage of HCl exposed specimens are respectively 9.43, 16.82 and 23.49%. So with the increment in rubber content, strength loss of acid-exposed specimen increases and this effect is considerably high when rubber is used without any treatment [20].

Youssf et al. [47] conducted a test for concrete surface abrasion. They made 16 disc specimens of RuGPC, each of dia 100 mm and 50 mm thick. They were given a number, dried for 24 hours at 100 °C temperature, and then the pretest weight of the specimen was calculated. These were clamped in the testing drum, which was filled with six hundred 16 mm diameter ball bearings (around 10 kg total) for the abrasion machine. The drum was mechanically rotated at a speed of 60 rpm for the test, which lasted 60 minutes. Each specimen was taken out once the test was finished; cleaned in a vaccum to avoid dust, and then the weight after the test was calculated. These measurements were used to calculate the loss of weight of each specimen and average it for each mix to determine the abrasion resistance. They noticed that as compressive strength diminishes, there is an increase in mass loss through abrasion. The abrasion of geopolymer concrete doubled when 20% of the amount of concrete sand was replaced with rubber. Again, this was a result of the rubber's poor adherence to the matrix, which made it simple for rubber particles to separate from the concrete surface that was being subjected to abrasion force [47].

7. Microstructures

Zhang et al. [41] compared the microstructures of geopolymer mortar specimens with crumb rubber prior to and following freeze-thaw cycles. SEM images of the control specimen before the freeze-thaw cycle indicate a dense and compact structure which indicates higher compressive strength. Cracks can be seen on the specimen after the freeze-thaw cycle on the SEM images which indicate lower compressive strength than the control specimen and improper geopolymerization [35, 41]. Mix with 10% crumb rubber content has a similar microstructure to the control specimen which was also compact and dense. This outcome also demonstrated that the morphology and microstructure of geopolymer pastes were not appreciably altered by the 10% substitution of river sand for crumb rubber.

Dong et al [14] displayed SEM images of dry rubber, rubber with water treatment, rubber treated with NaOH and activator-treated rubber. The SEM images show that the surface of the dry, untreated rubber particles was covered in microscopic particles, most likely made up of dust and grime. Due to the possibility that these tiny particles would separate the rubber from the geopolymer during mixing and trap air, adhesion would be decreased. After being adequately cleaned with water and NaOH, the surface texture retained small pores and fissures that could strengthen the bond with the concrete substrate. These findings from the SEM images support the possibility of high strength of geopolymer concrete with NaOH-treated rubber particles compared to the concrete with untreated rubber particles [14, 20].

8. Thermal Properties

Numerous factors, including aggregate used, density and the air volume in the mix, affect the thermal conductivity of concrete [30]. RuGPC mortar samples' thermal conductivity with a 100% replacement of fine aggregate with crumb rubber, according to Aslani et al. [23], was, on average, 79% lower than the control specimen. According to Zaetang et al. [24], thermal conductivity decreased by 82% when all fine aggregates were replaced with rubber crumbs, falling from 1.07 W/mK to 0.19 W/mK.. The reason for this drop is the high porosity and low density of the crumb rubber geopolymer specimen.

After two hours of exposure to high temperatures of up to 800°C, the mass loss and decline in compressive strength in RuGPC samples were examined by Luhar et al. [28]. A progressive mass loss was seen as the temperature increased, with higher mass loss observed for RuGPC samples than for GPC samples treated at the same temperature. RuGPC samples lost more compressive strength as temperature increased (52.4% at 800 °C vs. 45.2% at 800 °C), compared to GPC samples. The thermal conductivity also depends on the type of binder used. When the rubberized concrete mortar is compared with fly ash-based rubberized geopolymer mortar containing the same rubber percentage, Zaetang et al.[24] found a reduction of 35%.

9. Conclusions

The following conclusions have arrived from this work.

- The optimum value range of Molarity of NaOH used for the manufacture of RuGPC is 10-14M whereas the ratio of aluminosilicate base materials to alkali ranges from 0.4-0.6 and the dosage of superplasticizers varies from 1% to 2%.
- For RuGPC mixes, the early age strength is increased by heat curing whereas the rate of strength gain with ageing is higher for the ambient cured specimen.
- With higher CR replacement of aggregate, concrete slump/flow is reduced. A slump could be reduced by 50% with a 15% CR replacement as a result.
- When high CR replacement of aggregate is employed in RuGPC, the specimen's density decreases. This could lead to a density loss of 42% with 100% CR replacement.
- Water absorption and porosity of RuGPC specimen increase with crumb rubber content. Replacement of the control specimen with 100% crumb rubber increased the porosity by 1.5 times and the water absorption by 5.7 times.
- In the case of acid resistance of RuGPC, a compressive strength loss percentage of 23.49% observed with a 30% crumb rubber replacement.
- RuGPC often has a lower compressive strength than the respective control specimens. Compressive strength can be lowered by up to 60% when fine aggregates are substituted with 15% CR.
- Impact resistance increases with an increase in crumb rubber content. The number of blows for initial cracks and ultimate failure became 3 and 2.5 times that of the control specimen when 30% of fine aggregate replacement by crumb rubber.
- Thermal conductivity decreases with an increase in rubber content. With the replacement of 100% fine aggregates with crumb rubber, an 82% reduction in thermal conductivity was observed.

10. References

[1] Siddique R, Naik TR. Properties of concrete containing scrap-tire rubber - An overview. Waste Manag. 2004;24(6):563–9. 10.1016/j.wasman.2004.01.006
[2] Shen W, Shan L, Zhang T, Ma H, Cai Z, Shi H. Investigation on polymer-rubber aggregate modified porous concrete. Constr Build Mater 2013; 38:667–74. 10.1016/j.conbuildmat.2012.09.006
[3] Xiao Z, Pramanik A, Basak AK, Prakash C, Shankar S. Material recovery and recycling of waste tyres-A review. Clean Mater 2022;5.10.1016/j.clema.2022.100115
[4] S. Kaza, L. Yao, P. Bhada-Tata, F. Van Woerden, What a Waste 2.0: A Global Snapshot of Solid Waste Management to 2050, World Bank Publications, 2018.
[5] Sofi A. Effect of waste tyre rubber on mechanical and durability properties of concrete – A review. Ain Shams Eng J 2018; 9(4):2691–700. 10.1016/j.asej.2017.08.007
[6] P.K. Mehta, P.J. Monteiro, in Concrete: Microstructure, Properties, and Materials, McGraw-Hill, New York, 2005, p. 453.
[7] Meyer C. The greening of the concrete industry. Cem Concr Compos 2009;31(8):601–5.
[8] M.C. Juenger, F. Winnefeld, J.L. Provis, J.H. Ideker, Advances in alternative cementitious binders, Cem. Concr. Res. 41 (12) (2011) 1232–1243.
[9] Faridmehr I, Huseien GF, Baghban MH. Evaluation of mechanical and environmental properties of engineered alkali-activated green mortar. Materials (Basel). 2020;13(18).
[10] Lippiatt BC, Ahmad S. Measuring the Life-Cycle Environmental and Economic Performance of Concrete: the Bees Approach. International Workshop on Sustainable Development and Concrete Technology. 2004. 213–230 p.
[11] Singh PS, Trigg M, Burgar I, Bastow T. Geopolymer formation processes at room temperature studied by 29Si and 27Al MAS-NMR. Mater Sci Eng A. 2005;396(1–2):392–402.
[12] Shobeiri V, Bennett B, Xie T, Visintin P. A comprehensive assessment of the global warming potential of geopolymer concrete. J Clean Prod.2021;297:126669. 10.1016/j.jclepro.2021.126669
[13] McLellan BC, Williams RP, Lay J, Van Riessen A, Corder GD. Costs and carbon emissions for geopolymer pastes in comparison to ordinary portland cement. J Clean Prod. 2011;19(9–10):1080–90. 10.1016/j.jclepro.2011.02.010.
[14] Dong M, Elchalakani M, Karrech A, Yang B. Strength and durability of geopolymer concrete with high volume rubber replacement. Constr Build Mater 2021;274(xxxx):121783. 10.1016/j.conbuildmat.2020.121783.
[15] Pham TM, Liu J, Tran P, Pang VL, Shi F, Chen W, et al. Dynamic compressive properties of lightweight rubberized geopolymer concrete. Constr Build Mater 2020; 265:120753. 0.1016/j.conbuildmat.2020.120753.
[16] Arunkumar K, Muthukannan M, Suresh kumar A, Chithambar Ganesh A. Mitigation of waste rubber tire and waste wood ash by the production of rubberized low calcium waste wood ash based geopolymer concrete and influence of waste rubber fibre in setting properties and mechanical behavior. Environ Res 2021; 194 (December 2020):110661. 10.1016/j.envres.2020.110661.
[17] Gandoman M, Kokabi M. Sound barrier properties of sustainable waste rubber/geopolymer concretes. Iran Polym J (English Ed. 2015;24(2):105–12.
[18] Kumar S, Bishnoi M. The Fabrication of Geo-polymer Concrete using Rice husk, Crumb Rubber and GGBS. J Xidian Univ. 2020;14(9).
[19] Aly AM, El-Feky MS, Kohail M, Nasr ESAR. Performance of geopolymer concrete containing recycled rubber. Constr Build Mater 2019; 207:136–44.
[20] Saloni, Parveen, Pham TM, Lim YY, Malekzadeh M. Effect of pre-treatment methods of crumb rubber on strength, permeability and acid attack resistance of rubberised geopolymer concrete. J Build Eng. 2021;41(February):1–12.
[21] Lazorenko G, Kasprzhitskii A, Mischinenko V. Rubberized geopolymer composites: Effect of filler surface treatment. J Environ Chem Eng. 2021;9(4):105601.

[22] Singh B, Ishwarya G, Gupta M, Bhattacharyya SK. Geopolymer concrete: A review of some recent developments. Constr Build Mater [Internet]. 2015;85:78–90. Available from: http://dx.doi.org/10.1016/j.conbuildmat.2015.03.036.
[23] Aslani F, Deghani A, Asif Z. Development of Lightweight Rubberized Geopolymer Concrete by Using Polystyrene and Recycled Crumb-Rubber Aggregates. J Mater Civ Eng. 2020;32(2):04019345.
[24] Zaetang Y, Wongsa A, Chindaprasirt P, Sata V, Nakhon CS, Campus P, et al. Utilization of Crumb Rubber As Aggregate in High calcium fly ash geopolymer mortars. Int J GEOMATE. 2019;17(64):158–65.
[25] Mucsi G, Szenczi Á, Nagy S. Fiber reinforced geopolymer from synergetic utilization of fly ash and waste tire. J Clean Prod. 2018;178:429–40.
[26] Park Y, Abolmaali A, Kim YH, Ghahremannejad M. Compressive strength of fly ash-based geopolymer concrete with crumb rubber partially replacing sand. Constr Build Mater 2016;118(2016):44–51.
[27] Azmi AA, Abdullah MMAB, Ghazali CMR, Victor Sandu A, Hussin K. Effect of Crumb Rubber on Compressive Strength of Fly Ash Based Geopolymer Concrete. MATEC Web Conf. 2016;78:4–8.
[28] Luhar S, Chaudhary S, Luhar I. Thermal resistance of fly ash based rubberized geopolymer concrete. J Build Eng [Internet]. 2018;19:420–8. Available from: https://doi.org/10.1016/j.jobe.2018.05.025.
[29] Luhar S, Chaudhary S, Luhar I. Development of rubberized geopolymer concrete: Strength and durability studies. Constr Build Mater 2019;204:740–53.
[30] Wongsa A, Sata V, Nematollahi B, Sanjayan J, Chindaprasirt P. Mechanical and thermal properties of lightweight geopolymer mortar incorporating crumb rubber. J Clean Prod 2018;195:1069–80.
[31] Long WJ, Li HD, Wei JJ, Xing F, Han N. Sustainable use of recycled crumb rubbers in eco-friendly alkali activated slag mortar: Dynamic mechanical properties. J Clean Prod 2018;204:1004–15.
[32] Zhong H, Poon EW, Chen K, Zhang M. Engineering properties of crumb rubber alkali-activated mortar reinforced with recycled steel fibres. J Clean Prod 2019;238:117950.
[33] Dong W, Li W, Wang K, Vessalas K, Zhang S. Mechanical strength and self-sensing capacity of smart cementitious composite containing conductive rubber crumbs. J Intell Mater Syst Struct. 2020;31(10):1325–40.
[34] Valente M, Sambucci M, Chougan M, Hamidreza S. Reducing the emission of climate-altering substances in cementitious materials : A comparison between alkali-activated materials and Portland cement-based composites incorporating recycled tire rubber. J Clean Prod 2022;333(August 2021):130013.
[35] Si R, Guo S, Dai Q. Durability performance of rubberized mortar and concrete with NaOH-Solution treated rubber particles. Constr Build Mater 2017;153:496–505.
[36] Ameri F, Shoaei P, Reza Musaeei H, Alireza Zareei S, Cheah CB. Partial replacement of copper slag with treated crumb rubber aggregates in alkali-activated slag mortar. Constr Build Mater 2020;256:119468.
[37] Rashad AM, Sadek DM. Behavior of alkali-activated slag pastes blended with waste rubber powder under the effect of freeze/thaw cycles and severe sulfate attack. Constr Build Mater 2020;265:120716.
[38] Ali IM, Naje AS, Nasr MS. Eco-friendly chopped tire rubber as reinforcements in fly ash based geopolymer concrete. Glob Nest J. 2020;22(3):342–7.
[39] Yahya Z, Abdullah MMAB, Ramli SNH, Minciuna MG, Abd Razak R. Durability of Fly Ash Based Geopolymer Concrete Infilled with Rubber Crumb in Seawater Exposure. IOP Conf Ser Mater Sci Eng. 2018;374(1).
[40] Azrem Azmi A, Mustafa Al Bakri Abdullah M, Mohd Ruzaidi Ghazali C, Ahmad R, Musa L, Sheau Rou L. The Effect of Different Crumb Rubber Loading on the Properties of Fly Ash-Based Geopolymer Concrete. IOP Conf Ser Mater Sci Eng. 2019;551(1).

[41] Zhang B, Feng Y, Xie J, Lai D, Yu T, Huang D. Rubberized geopolymer concrete: Dependence of mechanical properties and freeze-thaw resistance on replacement ratio of crumb rubber. Constr Build Mater. 2021;310(October).
[42] 1. Toufigh V, Ghassemi P, Azizmohammadi M, Ghaemian M, Rajabi H, Ghassemi P. Mechanical properties and environmental impact of rubberized fly ash- and red mud-based geopolymer concrete. Eur J Environ Civ Eng 2021;0(0):1–24.
[43] Rajendran M, Akasi M. Performance of Crumb Rubber and Nano Fly Ash Based Ferro-Geopolymer Panels under Impact Load. KSCE J Civ Eng. 2020;24(6):1810–20.
[44] Montes C, Zang D, Allouche EN. Rheological behavior of fly ash-based geopolymers with the addition of superplasticizers. J Sustain Cem Mater. 2012;1(4):179–85.
[45] Luhar S, Chaudhary S, Dave U. Effect of different parameters on the compressive strength of rubberized geopolymer concrete. Multi-disciplinary Sustain Eng Curr Futur Trends. 2016;77–86.
[46] Dong M, Elchalakani M, Karrech A. Development of high strength one-part geopolymer mortar using sodium metasilicate. Constr Build Mater 2020;236:117611.
[47] Youssf O, Elchalakani M, Hassanli R, Roychand R, Zhuge Y, Gravina RJ, et al. Mechanical performance and durability of geopolymer lightweight rubber concrete. J Build Eng 2022;45.47
[48] Erdem S, Dawson AR, Thom NH. Microstructure-linked strength properties and impact response of conventional and recycled concrete reinforced with steel and synthetic macro fibres. Constr Build Mater. 2011;25(10):4025–36

IOP Conf. Series: Earth and Environmental Science 1149 (2023) 012010 doi:10.1088/1755-1315/1149/1/012010

Study on the effectiveness of prefabricated cage system reinforcement in columns

T Thejarathnam[1], M H Prashanth[2], Imran Kuttagola[3]

[1] Former PG Student, NITK, Surathkal, Mangalore 575025, India
[2] Assistant Professor, NITK, Surathkal, Mangalore 575025, India
[3] Research Scholar, NITK, Surathkal, Mangalore 575025, India

imrank.207cv006@nitk.edu.in

Abstract. The concrete is the primary vertical load bearing component in a reinforced concrete column whereas the steel cage provides additional vertical load carrying capacity along with the confinement of core concrete. The research paper will focus on the performance of Prefabricated Cage System (PCS) Reinforcement in square columns. Eight columns were casted and tested, out of which two are rebar reinforced columns and six are PCS reinforced columns. There were two test groups, each group containing one rebar column and three PCS columns. In this study work is done to compare the performance of a PCS reinforced column to a rebar reinforced column, in addition comparing PCS cages of differently sized grid openings. The objective is to theoretically, experimentally, and numerically investigate the PCS column for its load carrying capacity and displacement capacity. From the results it is observed that as the size of opening of a PCS reinforced column increase, the load carrying capacity increases as the concrete running through the openings can strengthen the connection. Hence to improve the effectiveness of the PCS columns the spacing of openings should be increased. Further, the stiffness and energy absorption are more in rebar reinforced column compared to all PCS columns.

1. Introduction

1.1. General
Concrete is basically a heterogeneous mixture of cement, coarse aggregate, fine aggregate, and water with some voids. As the concrete is weak in tension, steel is used as the reinforcement material in areas where high tensile stresses are to be compensated. The steel and concrete combination is reinforced concrete system which is a composite section and is popularly used as construction material for structural and non-structural members. The rebar reinforcement system consists of two types of bars. One is the longitudinal bar which resist compressive or tensile loads, and the other is the transverse bar which helps to confine the concrete inside the reinforcement (core concrete). The core concrete and the longitudinal steel are the main carriers of the compressive load in a RC column. The transverse bars are bent to fit around the longitudinal bars and then tied into place firmly [1, 2].

1.2. Proposed Prefabricated Cage System (PCS)
A new innovative reinforcement system called as Prefabricated Cage System reinforcement (PCS) was introduced by Halil Sezen. In this system, the total reinforcement with longitudinal and transverse are monolithically prefabricated using a hollow steel tube. A Prefabricated Cage System (PCS)

reinforcement can be fabricated by cutting out the rectangular openings on a steel plate with a continuous weld along the edge. The vertical continuous strips act as longitudinal reinforcement and the horizontal strips act as transverse reinforcement and thereby acting monolithically [1]. More than one PCS reinforcement layers can be placed concentrically in a member if higher reinforcement ratios are required. Connecting two PCS cages with different sizes can be done easily. For example, a PCS reinforcement can slide into another PCS reinforcement as deep as required for development length. The concrete running through the openings can strengthen the connection. In this case, the size and spacing of the openings of the two PCS cages can be different to provide the strongest bond between the concrete and the PCS reinforcement. The opening dimensions, their spacing, and whether it is better or not to have offset openings, can be investigated through experimental testing [2].

The proposed system of reinforcement is expected to be an alternative for the existing reinforcement system for the structural members mainly beams and columns. Prefabricated Cage System (PCS) reinforcement can be very effective in columns as the confinement capabilities of PCS reinforcement are higher than that of rebar reinforcement. PCS has major advantages over conventional rebar reinforced system. Perfectly uniform transverse steel spacing can be achieved by this system. This system eliminates the labor associated with cutting, bending, and tying reinforcing bars and reduce the construction time and cost of construction. Also, cage can be placed easily in PCS with no steel congestions problems in heavily reinforced sections [3].

Important investigations concerned with prefabricated cage system reinforcement and concrete confinement models have been carried out and contributed to the literature. Mander et al. proposed a single expression to represent the entire stress strain curve for confined reinforced concrete [4]. Hoshikuma et al. proposed two expressions to represent the stress strain curve, one for ascending limb and other for descending limb which provides the better agreement with the test results [5]. Shamsai et al. proposed a new reinforcement system called Prefabricated Cage system Reinforcement (PCS) in which a solid steel tube is made to act as both longitudinal and transverse reinforcement and the results show that the confinement capacity of PCS column is more than the rebar column resulting in higher strength and displacement capacity [2]. Chithra et al. studied the flexural ductility of Prefabricated Cage reinforced steel-concrete composite beams and concluded that the ductility and energy absorption capacity of PCRC beams are more compared to PCS beams [3].

Though, investigations have been done on prefabricated cage system reinforcement, a research gap is found in the literature regarding the effectiveness of PCS cages of differently sized grid openings. Hence, effort has been made in the current work to do an extensive study on the effectiveness of prefabricated cage system reinforcement with various sizes of openings in columns. Further, with the present research work on PCS reinforcement, this system can be used as a best alternative for the existing rebar reinforcement system for the structural members, mainly beams and columns and can reduce the construction time and cost of construction. Hence the main objective of this paper is to study the behaviour of small-scale column specimens of PCS reinforcement with different sizes of openings and to compare the same with rebar reinforcement specimens by experimental investigation, analytical confinement models, and finite element modelling in ABAQUS.

2. Experimental, Analytical and Numerical program

2.1. Materials

Ordinary Portland cement of 53 grade is used. It is tested as per the guidelines provided in the code IS: 4031-1968 and IS: 8112-1989 recommendations for the hydraulic cement. The test results of cement are shown in table 1, which confirms the requirement as per the IS Code. Locally available natural sand passing through 4.75 mm sieve, confirming to zone II of IS 383:1970 is made use. The coarse aggregates with the maximum size of 20mm and the maximum size of 12.5 mm are used in 80:20 proportion respectively by weight. The physical properties of fine aggregates and coarse aggregates were determined and are listed in the table 2. High yield strength deformed bars (HYSD) designated as Fe415 has been used for reinforcement. The steel was tested in tension and the yield stress is obtained as 417.23

IOP Conf. Series: Earth and Environmental Science 1149 (2023) 012010 doi:10.1088/1755-1315/1149/1/012010

MPa. The final mix proportion as per the guidelines for concrete mix design provided in the code IS: 10262-1982 is shown in table 3 [6]. The 28 days compressive strength for the hardened concrete mixes tested on 150 mm x 150 mm cubes is shown in table 4.

Table 1. Test results of Cement.

Test Parameters	Results	As per IS 12269-2013 (Specifications of 53 grade OPC cement)
Initial Setting Time	49 minutes	Not less than 30 minutes
Final setting time	357 minutes	Not more than 600 minutes
Specific gravity	3.12	-
Soundness	2 mm	Less than 10 mm
Normal Consistency	30%	-

Table 2. Physical properties of fine aggregates and coarse aggregates.

Property	Fine aggregate Value	Coarse aggregate Value
Specific Gravity	2.6	2.6
Moisture content	Nil	Nil
Water Absorption	1%	0.5%

Table 3. Final mix proportion.

Ingredients	Quantity
Cement	400.55 kg/m^3
Fine Aggregate	701.039 kg/m^3
Coarse Aggregate	1096.51 kg/m^3
Water	180.25 kg/m^3
Mix Proportion Ratio (C : FA : CA : W)	1 : 1.75 : 2.74 : 0.45

Table 4. The 28 days compressive strength for the hardened concrete mixes.

Column	Cubes (150 x 150 mm)	Ultimate Load (kN)	f_{ck} (N/mm^2)	Average f_{ck} (N/mm^2)
	1	840	37.33	
Re20	2	810	36.00	36.29
	3	800	35.56	
	1	860	38.22	
PCS180	2	830	36.89	36.89
	3	800	35.56	
	1	880	39.11	
PCS150	2	850	37.78	38.51
	3	870	38.67	
	1	830	36.89	
PCS190	2	840	37.33	37.78
	3	880	39.11	

2.2. Reinforcement details

Two test groups, each group containing one rebar column and three PCS columns, with a total of eight columns are prepared and tested to investigate the performance of PCS reinforcement and Rebar reinforcement columns. Among the eight columns two columns are rebar reinforced and remaining six are PCS reinforced columns. It is ensured that the volumes of rebar reinforcement and PCS reinforcement are equal. The buckling and other problems associated with slender columns are avoided, by designing the short columns in the experiments. Thickness of plate in PCS reinforcement is kept uniform in all the PCS specimens in order to maintain equal volume of reinforcement and at the same time keeping the different sizes of the opening. The design of rebar reinforced columns has been done and the longitudinal reinforcement is provided with 4 number of 20 mm diameter bars of Fe415 grade steel and transverse reinforcement is provided with 8mm diameter lateral ties @100 mm c/c spacing [7]. For PCS columns, the uniform thickness of plate in every column is 4 mm and the effective cover being 25mm. Grade of steel Fe250 has been used in the cage reinforcement. Size of the openings provided are 180 × 60 mm, 150 × 72 mm, and 190 × 57 mm. The experimental test matrix is shown in table 5. The reinforcement details of Rebar and PCS are shown in figure 1.

Table 5. Experimental test matrix.

Column type	Reinforcement Type	Tube Thickness (mm)	#Bars	Opening Dimension (mm)
PCS180	PCS	4	4	180 × 60
PCS150	PCS	4	4	150 × 72
PCS190	PCS	4	4	190 × 57
RE20	REBAR	4	4	-

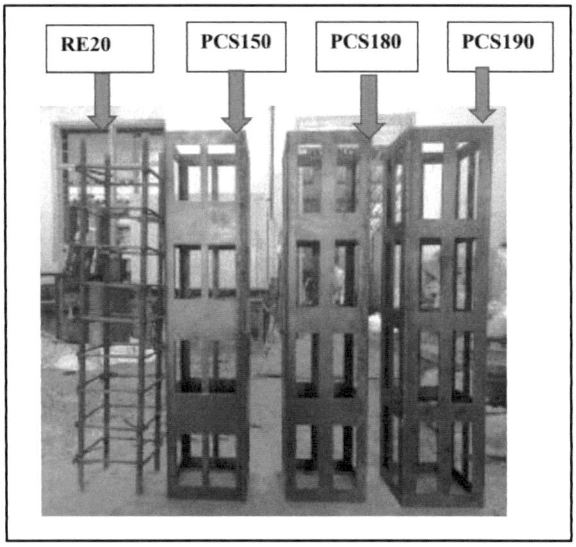

Figure 1. Reinforcement details of Rebar and PCS.

Figure 2. Column tested in compression frame bed.

2.3. Experimental setup and Testing of Columns

PCS reinforcement and Rebar reinforcement columns are tested in compression frame bed. The experimental setup is shown in figure 2. All the eight columns were tested under load control for compression. Proving ring division is noted corresponding to load from calibration chart. Load is applied at an increment of 72 kN and axial deflections are measured from the dial gauges. At each load increment, columns were explored carefully for cracks with the help of magnifying lens which they were marked using a black pencil. Experimental data was recorded and tabulated. The columns were loaded till failure and the mode of failure of the specimens were observed. After failure of columns, the cracks were marked with the blue marker pen and photo was captured.

2.4. Analytical confinement models

A confinement model proposed by Bouafia et al. is used to obtain the theoretical stress-strain plots [8]. This confinement model is made by modifying the existing Mander's model. The stress-strain curve of confined concrete can be determined by three coordinates:

- The coordinate (ε_{cc0}, f_{cc}) corresponding to the peak stress-strain
- (ε_{65}, $0.65f_{cc}$) corresponding to the representative point of the stress – strain curve at 0.65fcc after the peak and
- (ε_{ccu}, $0.65f_{cc}$) corresponding to the ultimate strain as shown in the figure 3.

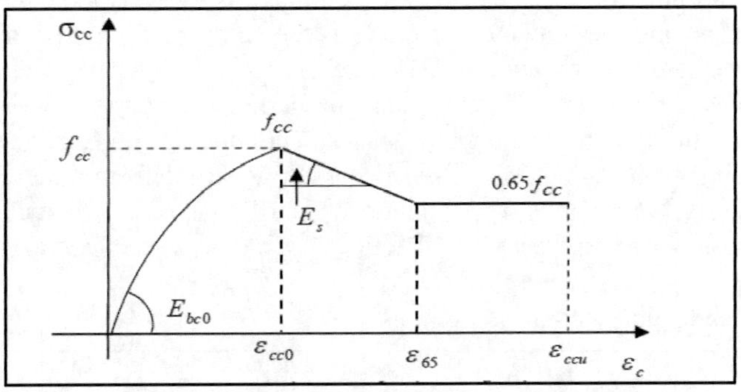

Figure 3. Stress-strain curve for the confined concrete.

The stress-strain relation in the ascending region between the zero and the first coordinate,

$$\sigma_{cc} = f_{cc} \times \frac{kc \times \left(\frac{\varepsilon_{cc}}{\varepsilon_{cco}}\right) + (k'_c - 1) \times \left(\frac{\varepsilon_{cc}}{\varepsilon_{cco}}\right)^2}{1 + (k_c - 2) \times \left(\frac{\varepsilon_{cc}}{\varepsilon_{cco}}\right) + k'_c \times \left(\frac{\varepsilon_{cc}}{\varepsilon_{cco}}\right)^2} \qquad \text{For: } 0 \le \varepsilon_c \le \varepsilon_{cc0} \tag{1}$$

Where $\varepsilon_{cc0} = \varepsilon_{c0} \times \left[1 + 5\left(\frac{f_{cc}}{f_{co}} - 1\right)\right]$ $k_c = \frac{E_{bco} + \varepsilon_{cco}}{f_{cc}}$ $E_{bc0} = 11000 \times \sqrt[3]{f_{cc}}$ $k'_c = k_c - 1$

The stress-strain relation of the descending region between the first and the second coordinates,

$$\sigma_{cc} = f_{cc} - E_s(\varepsilon_c - \varepsilon_{cc0}) \qquad \text{For: } \varepsilon_{cc0} < \varepsilon_c \le \varepsilon_{65} \tag{2}$$

Where E_s is the slope of the descending curve and can be defined as:

$$E_s = \frac{6 \times {f_{c0}}^2}{k_e \times \rho_s \times f_{yh}} \tag{3}$$

And ε_{65} is the strain corresponding to the stress equal $0.65f_{cc}$. It can be defined by:

$$\varepsilon_{65} = \varepsilon_{cc0} + \frac{0.35f_{cc}}{E_s} \tag{4}$$

After reaching the stress $0.65f_{cc}$, the stress of the confined concrete is a constant value 0.65fcc regardless of the increasing strain until the ultimate strain (ε_{ccu}). The ultimate strain of confined concrete adopted in this study can be defined by taking account the ultimate strain of transverse reinforcement which is given as:

$$\varepsilon_{ccu} = \varepsilon_{cu} + 0.4 \times \frac{f_l}{f_{co}} \quad = 0.0035 + 0.4 \times \frac{f_l}{f_{co}} \tag{5}$$

Where f_l is the lateral confining stress and f_{c0} is the unconfined concrete compression stress.

2.5. Modelling using ABAQUS

The analysis of PCS and Rebar columns are modelled and analysed using ABAQUS software. This finite element modelling comprises of an idealized linear brick C3D8R element type and meshing of elements to replicate the column of size 1000×250×250 mm. The numbering of nodes follows the convention as shown in figure 4. A 3-D, deformable, solid part is used to model the column. Embedded interaction has been used and reinforcement is made as the embedded region and concrete is made as the hosting region. The assembly module in ABAQUS is shown in figure 5. The analysis of PCS and Rebar columns includes non-linear concrete and steel properties which are shown in table 6. Column is axially loaded with the simulation done under displacement control, ranged in magnitude from 0 to 5 mm, with one end being encastred and the other being allowed to deform under axial deformation as shown in the figure 6. Meshing is done by parts independently to the concrete and steel separately with a mesh size of 25 mm as shown in figure 7. Full analysis job type with ODB format is created and the results are obtained.

Table 6. Concrete and steel properties.

Property	Concrete	Steel
Compressive strength of concrete (28 days)/ Characteristic strength of steel (MPa)	25	250
Modulus of elasticity (E) (MPa)	25000	2×10^5
Mass density (ton/mm^3)	2.4×10^{-9}	7.85×10^{-9}
Poisson's ratio	0.15	0.31

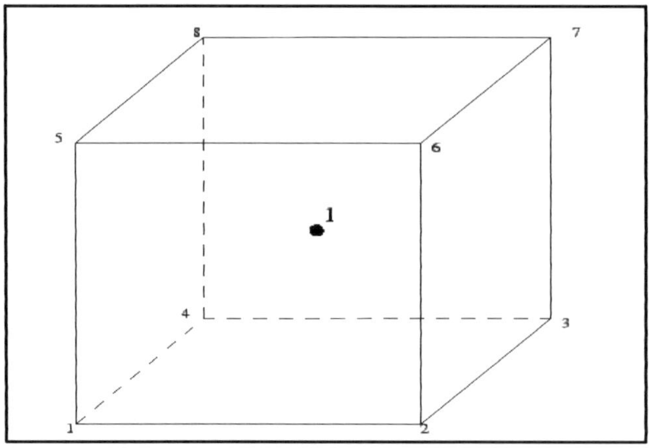

Figure 4. An 8-node linear brick element.

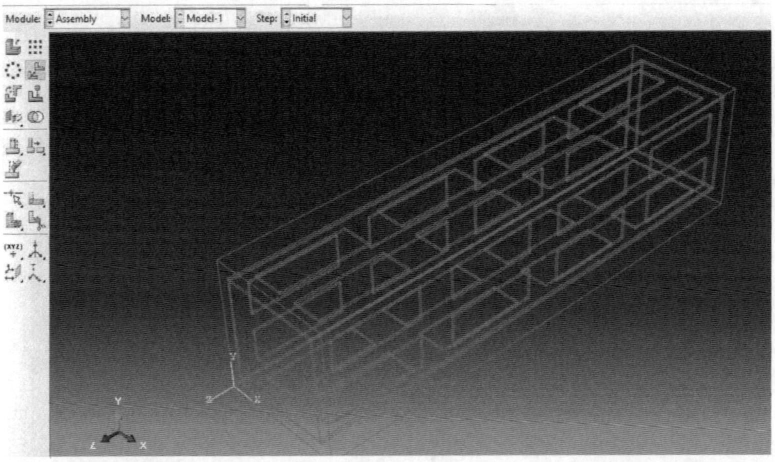

Figure 5. Assembly module in ABAQUS.

Figure 6. Loads and Boundary Conditions.

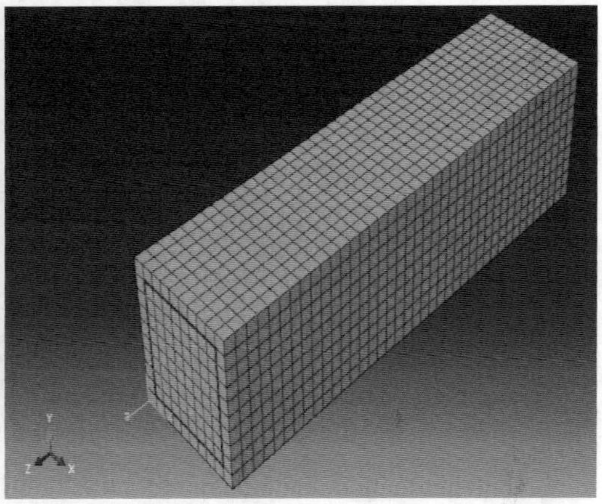

Figure 7. Mesh Module in ABAQUS.

3. Results and discussions

3.1. Experimental Results and Discussions

The experimental results obtained from the testing of eight columns have been presented in the form of figures and tables. The results have been plotted for load versus deformation. The comparison graphs between the columns have been plotted for the performance evaluation. The load versus deformation curves obtained are shown in figure 8. The slope of load versus deformation has been observed to be similar in all the specimens. The peak load is more in RE20 followed by PCS190, PCS180, and PCS150. Hence the load carrying capacity of a rebar reinforced column is more compared to a PCS reinforced column. Further, as the size of opening of a PCS reinforced column increase, the load carrying capacity increases as the concrete running through the openings can strengthen the connection. Axial deformations at failure are same in all the columns as mentioned in table 7. Stiffness and energy absorption is more in RE20 compared to all the PCS columns. Among PCS columns stiffness is more and energy absorbed is less in PCS150. In PCS180 energy absorbed is more but the stiffness is less.

Table 7. Experimental results.

Specimen Designation	First crack load (kN)	Failure load (kN)	Failure deformation (mm)	Stiffness (kN/mm)	Energy absorbed (kN-mm)
RE20	700	1018.20	1.20	826.73	496.04
PCS150	600	654.70	1.01	703.35	355.19
PCS180	600	763.60	1.38	609.16	420.93
PCS190	700	872.73	1.22	636.12	389.31

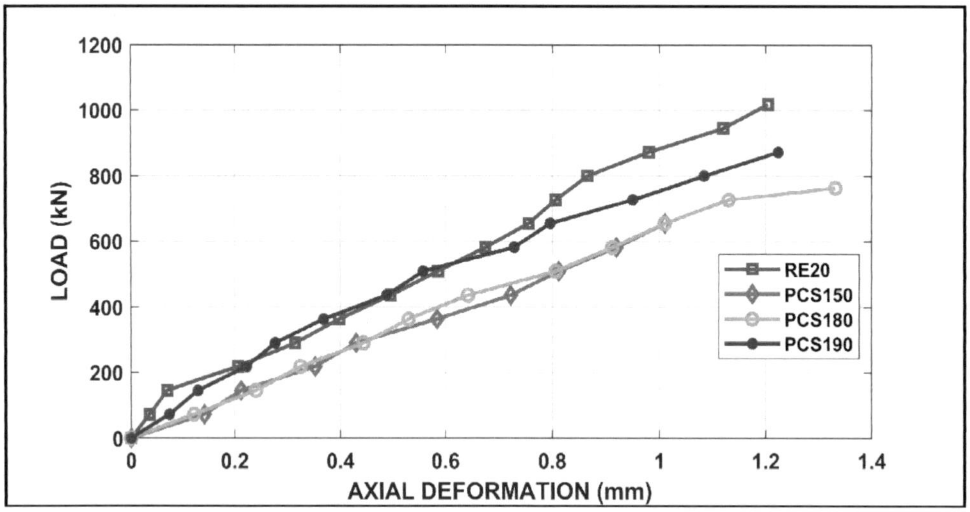

Figure 8. Load versus deformation plot for all columns.

Crack patterns have been observed for all columns as shown in figure 9. Cracks observed were very wide in PCS150 as the longitudinal cage reinforcement failed earlier, but the specimens RE20, PCS 190 have shown very less cracks with shorter propagation depths. Spalling of cover concrete is also more in PCS150. This is because of the bulging of longitudinal cage reinforcement. In all the PCS reinforced columns, spalling of cover concrete occurred quickly after the initiation of cracking due to de-bonding between concrete and cage reinforcement.

(a) Crack pattern in PCS180.

(b) Crack pattern in PCS150.

(c) Crack pattern in PCS190.

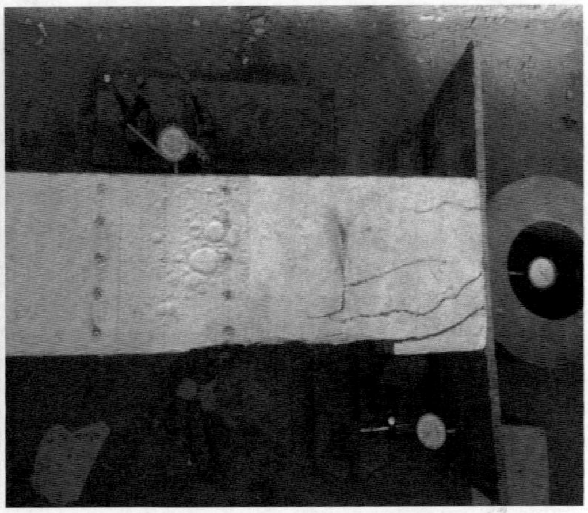

(d) Crack pattern in RE20.

Figure 9. Crack patterns for all columns.

3.2. Analytical Results and Discussions

According to the Boufia's model, PCS150 performs better than all specimens. It is because the confinement effect has played an important role due to the amount of transverse reinforcement provided. Also, it is obvious that the transverse cage reinforcement is more in PCS150, the performance is better. In the ascending limb, the confined compressive strength (f_{cc}) is more in PCS150 because the confinement mainly depends on the area of transverse reinforcement provided. In the descending limb i.e., softening portion, depends on confined concrete strain (ε_{cco}) which is function of (f_{cc}). Hence in this theoretical model, the quantity of the transverse reinforcement provided decides the confinement of concrete and it is difficult to arrive at the actual load versus deformation behaviour of an axially loaded column. Stress-strain plot for all columns derived by the Boufia's model is shown in figure 10.

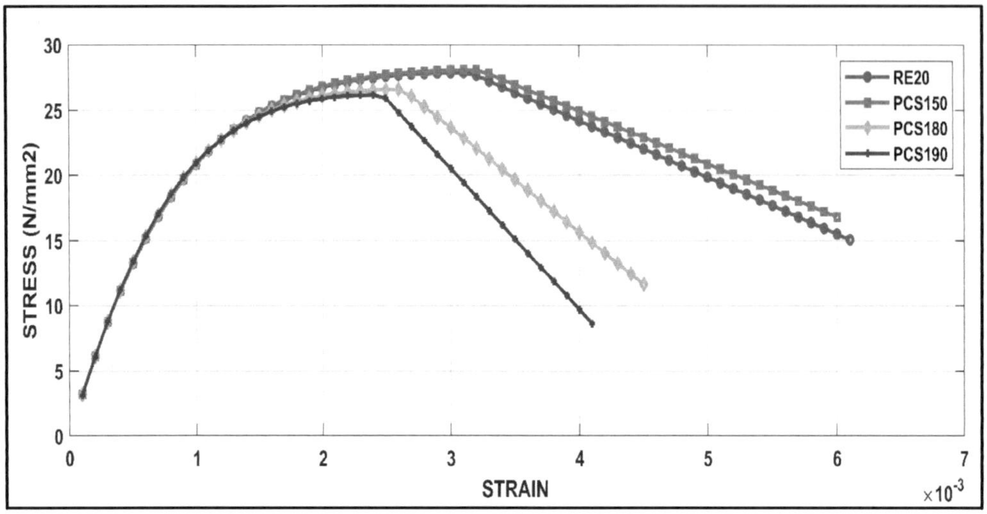

Figure 10. Stress versus strain plot for all columns.

3.3. Numerical Results and Discussions

Numerical Results of columns from Abaqus (shown in table 8) are taken for single node using unique nodal system in the software. Total number of nodes at a section were 121. At a particular node, the peak load has been compared for all the columns. Hence from the FEM analysis, the peak load of rebar specimen is having a slightly higher peak load as shown in figure 11. From the stress contours (figure 12) and load versus deformation plots, all the specimens achieved their peak load at around 2 mm of axial deformation. But, a drop in load was observed in the PCS150 and RE20 specimens after the peak load and not in the other specimens. The reason behind the drop in the load is due to the buckling of longitudinal reinforcement. The energy absorbed is more in PCS180 and PCS190 compared to PCS150. Ultimate load of RE20, PCS180 and PCS190 are more than PCS150 in the softening part as shown in figure 11. The deformation trend after the peak load is almost same for PCS180, PCS190 and RE20. Hence to improve the effectiveness of the PCS columns the spacing of openings should be increased, so that the confinement increases.

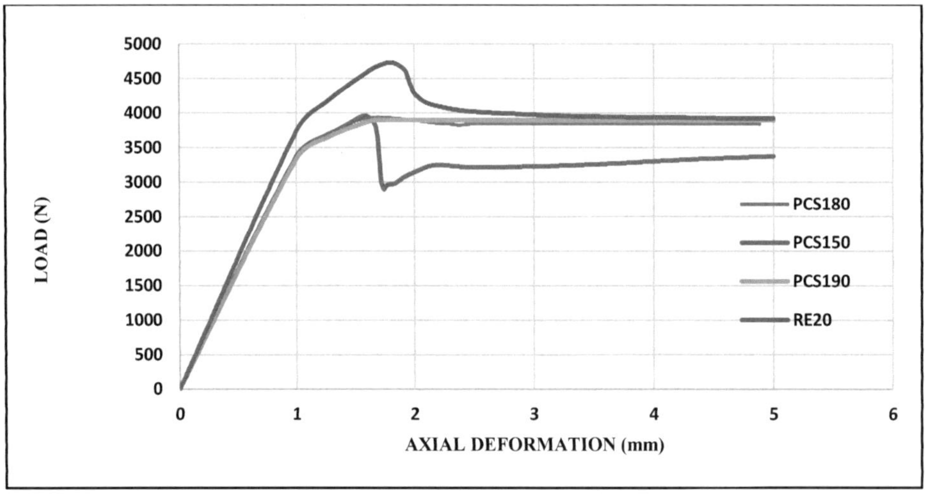

Figure 11. Load versus axial deformation of all columns from Abaqus.

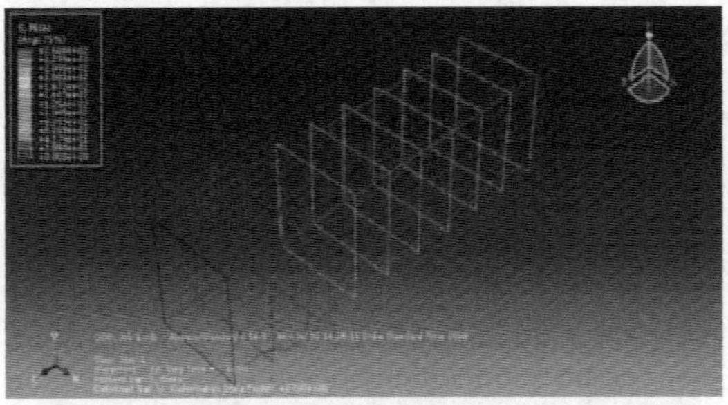

(a) Stress contour of reinforcement of RE20

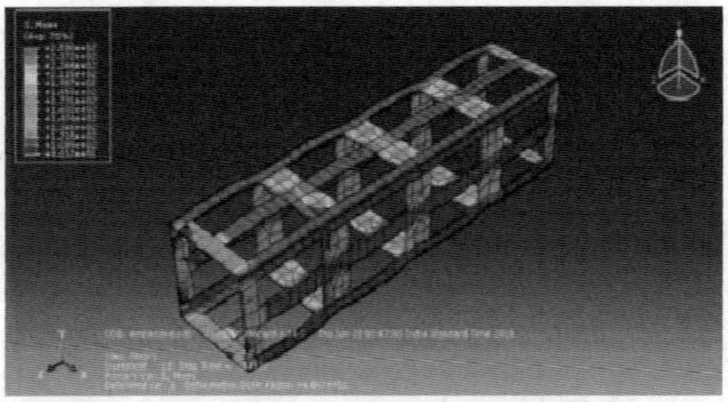

(b) Stress contour of reinforcement of PCS180

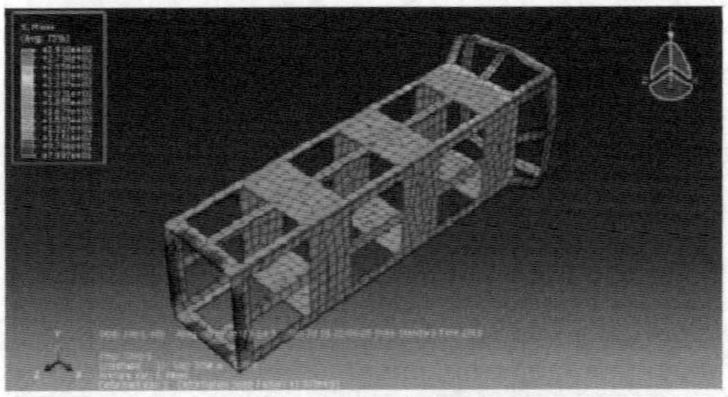

(c) Stress contour of reinforcement of PCS150

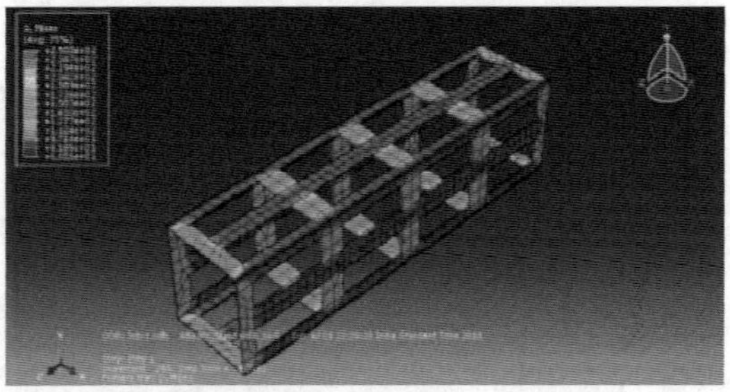

(d) Stress contour of reinforcement of PCS190

Figure 12. Stress contour of reinforcement for all columns.

Table 8. Numerical Results of columns from Abaqus.

Specimen	Peak load (N)	Ultimate Load (N)	Peak load deformation (mm)
RE20	4732	3924	1.80
PCS150	3959	3373	1.60
PCS180	3962	3839	1.63
PCS190	3954	3896	1.76

4. Conclusions

Based on the experimental, analytical, and numerical results obtained, the following conclusions can be drawn.

- From the experimental results, it can be observed that the peak load is more in RE20 than the PCS columns as the size of openings is more in RE20 followed by PCS190, PCS180, and PCS150, due to which the concrete running through the large openings of RE20 can strengthen the connection. Hence the load carrying capacity of a rebar reinforced column is more compared to a PCS reinforced column.
- Further, it can be observed that the stiffness and energy absorption is more in RE20 compared to all PCS columns till the peak load. (No post peak response is obtained as the experimental tests are load controlled).

- Cracks observed were very wide in PCS150 as the concrete running through the small openings of PCS150 is less and the longitudinal cage reinforcement failed earlier due to weak connection and de-bonding between concrete and cage reinforcement, but the specimens RE20, PCS 190 have shown very less cracks with shorter propagation depths as the concrete running through the large openings can strengthen the connection and sufficient bond is developed between concrete and cage reinforcement.
- Spalling of cover concrete is also more in PCS150 due to large bulging of longitudinal cage reinforcement.
- From the analytical results, PCS150 performed well compared to all other columns due to more transverse cage reinforcement. Hence based on the result of PCS150, it will be easier to decide the amount of transverse reinforcement required to achieve the load carrying capacity.
- From the numerical results, PCS180 and PCS190 performed well when compared to RE20 and PCS150 (where the drop was observed). The post peak response of PCS 180 and PCS 190 is like that of RE20.
- For the same volume of reinforcement, the configuration of PCS reinforcement must be changed to make the performance as better as rebar column as the study shows that the performance of the PCS columns depends on the size of openings and the spacings of the openings and by adopting the best configuration of PCS reinforcement, the performance can be improved.

References

[1] Shamsai M and Sezen H 2011 Behavior of square concrete columns reinforced with prefabricated cage system *Materials and Structures* **44(1)** pp 89–100
[2] Shamsai M and Sezen H 2005 Fast and easy concrete construction using innovative steel reinforcement *Proc. in Construction Research Congress (San Diego, California)* pp 1–10
[3] Chithra R, Thenmozhi R and Ravathi M C 2011 Flexural ductility of prefabricated cage reinforced steel-concrete composite beams *Asian Journal of Civil Engineering (Building and Housing)* **12(6)** pp 719–29
[4] Mander J B, Priestley M J and Park R 1988 Theoretical stress-strain model for confined concrete *Journal of Structural Engineering* **114(8)** pp 1804–26
[5] Hoshikuma J, Kawashima K, Nagaya K and Taylor A W 1997 Stress-strain model for confined reinforced concrete in bridge piers *Journal of Structural Engineering* **123(5)** pp 624–33
[6] IS Standard *Recommended guidelines for concrete mix design* 2004 pp *10262-1982*
[7] Devadas M 2003 *Reinforced concrete design* vol 2 (Tata McGraw-Hill Education)
[8] Bouafia Y, Iddir A, Kachi M S and Dumontet H 2002 Stress–strain relationship for the confined concrete *Material Science Engineering* pp 70-80

IOP Conf. Series: Earth and Environmental Science 1149 (2023) 012011 doi:10.1088/1755-1315/1149/1/012011

Numerical investigation of the hybrid reinforced concrete beam using GFRP bars

Aditya Kumar[1], M H Prashanth[2], Imran Kuttagola[3]

[1] M Tech (Research), NITK, Surathkal, Mangalore 575025, India
[2] Assistant Professor, NITK, Surathkal, Mangalore 575025, India
[3] Research Scholar, NITK, Surathkal, Mangalore 575025, India

imrank.207cv006@nitk.edu.in

Abstract. The numerical study of the hybrid reinforced concrete beam using GFRP bars has been explored in this research. Finite element software ABAQUS had been used for numerical investigation. In the present research, the use of GFRP bars as a replacement to steel reinforcing bars for longitudinal reinforcement is attempted. Further, a RC beam by using GFRP as longitudinal reinforcement and steel rebars as transverse reinforcement is compared with the conventional steel reinforced concrete beam. The beams are tested for three-point bending using numerical simulation under displacement control. By performing the numerical modelling, load, strain and displacement data has been attained. Further, by using the output data, stress-strain curves of concrete, load versus strain variations in tension bars and load versus deflection curves are plotted. The stresses, deflections, strain developed in longitudinal bars, load carrying capacity, energy absorption are then calculated using the data and are compared. From the results it is observed that, the load carrying capacity decreases by 16% and the energy absorption increases by 29% when longitudinal steel bars are replaced by GFRP bars.

1. Introduction

During the last few decades, the use of contemporary Fibre Reinforced Polymer (FRP) bars for strengthening and repairing structural elements such as beams, columns and slabs has been increasing. Fibre reinforced composites have been employed in construction applications both for new installation and rehabilitation/refurbishment of existing infrastructures. For RC structural members, FRP as a reinforcement is a better choice with comparison to steel because it provides higher strength to weight ratio. Mostly four types of FRP are employed for structural support: AFRP (Aramid Fibre Reinforced Polymers), CFRP (Carbon Fibre Reinforced Polymers), BFRP (Basalt Fibre Reinforced Polymers), and GFRP (Glass Fibre Reinforced Polymers) [1]. According to some studies, the reinforcing bar configurations are critical in enhancing the compressive and flexural strength for RC beam reinforced with FRP bars. Many researchers have devoted a lot of time and money to improve the effectiveness of concrete with FRP bars. Besides this, RC members exposed to coastal environment, the reinforced component's endurance is a serious problem. Corrosion of steel reinforcements in structural components is a common source of mechanical weakness in RC members. FRP rebar, especially GFRP, has been found to be high strength to weight ratio and are being non-conductive, thereby reducing the chances to corrosion [3].

2. Literature review

Panahi and Izadinia studied on strengthening of beams with mounted FRP rods using ABAQUS software. The mid-span maximum displacement, breakdown deformation, as well as ductile indices and numerical findings are summarized. At the end of this study, the effectiveness of NSM (Near Surface Mounted) FRP bars in reinforcing steel and concrete beams has been validated [2]. Hamad et al. conducted theoretical investigation on the slippage impacts of FRP and steel rebars, on pre-heating and post-heating, over the behaviour of reinforced concrete beams. In order to predict the load capacities of RC beams, a new theoretical method was proposed with consideration of slippage between rebars and surrounded concrete. Further, they concluded that with increase in temperature, the beams containing steel bars demonstrated pure torsional breakdown, as well as a large number of fractures and smaller average break widths than equivalent beams using FRP bars [3]. Ge et al. presented the experimental findings of fibre reinforced concrete hybrid beams using Engineered Cementitious Composite (ECC) concrete. ECC thickness, reinforcement type and quantity of reinforcement were the main parameters of the study. In their study four ECC beams were studied: reinforced with steel, FRP, mix of steel and FRP bars, and a higher amount of only FRP reinforcement. They concluded that ECC layer in tension zone of RC beam improves the stiffness and ultimate moment capacity of RC beam. Also, it was observed that the ability to resist cracks and deformation in RC beams can improved by using ECC layer in RC beams [4]. Protchenko and Szmigiera experimentally investigated fire resistance properties of reinforced beams with hybrid bars. They found that, when exposed to increased temperatures and even during the freezing process, the behaviour of FRP RC beams appeared extremely unusual [5]. Hekal et al. suggested various U-shaped, vertical strips, one-sided and two-sided strengthening techniques for reinforced concrete T-beams which were analysed by Finite element method. In order to improve the shear capacity, Ultra High Performance Fibre Reinforced Concrete (UHPFRC) plates are affixed to reinforced T-beams using epoxy adhesive. They suggested that U-shaped technique is the best strengthening technique for reinforced concrete T-beams [6].

At site, for the fabrication of reinforcement, it is required to bend the reinforcement bars, particularly for stirrups in RC members. When GFRP bars are used, there is difficulty in bending of these bars due to less bendability (less percentage of elongation). Hence to overcome this drawback of GFRP bars, hybrid RC beam is employed with steel bars as stirrups and FRP as longitudinal bars. Hence the novelty for the present work is developing a hybrid beam using GFRP bars as longitudinal bars and steel bars as stirrups.

3. Objective

The objective of this paper is to present herewith the numerical investigation of the behaviour of RC beam with GFRP bars and steel bars. To model and analyze the behaviour of reinforced concrete beams, the following cases have been considered:

- Conventional RC Beams using steel for both longitudinal bars and stirrups
- RC beam using GFRP as longitudinal bars and steel bars as stirrups

4. Methodology

The RC beam is tested for three-point bending using numerical simulation. The beam is simply supported across a span of 2000 mm length and having cross-section of 100 mm x 200 mm (figure 1). The study is conducted for the two cases of longitudinal reinforcement i.e., steel rebars and GFRP bars. In each of the case, the bottom tension reinforcement is provided with 2 numbers of 12 mm diameter bars and top reinforcement is provided with 2 numbers of 12 mm diameter hanger bars. For both the beams, the stirrups are provided with transverse 2-legged HYSD bars of 6 mm diameter at spacing of 100 mm.

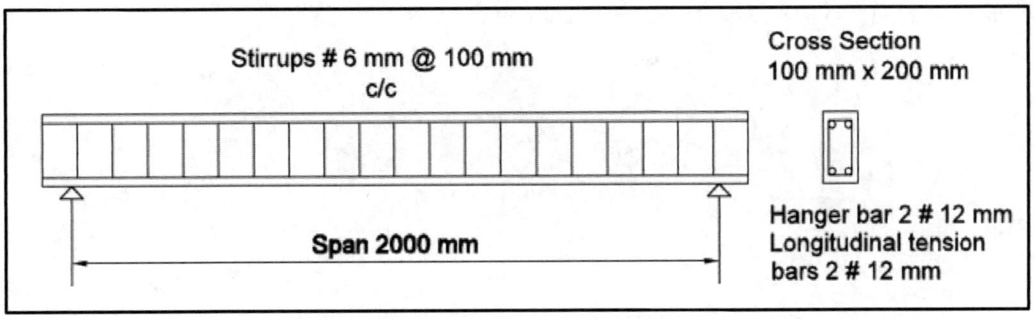

Figure 1. The details of RC beam.

Table 1. Material properties of deformed steel bars and GFRP bars.

Properties	Steel	GFRP
Density (gm/cm^3)	7.85	2.50
Tensile Strength (MPa)	415	3450
Young's Modulus (GPa)	200	72
Elongation (%)	8.0	4.0
Coefficient of Linear Expansion (10^{-6} /°C)	11.7	8.0

The various parts of the RCC are modelled individually with their own types. Concrete is modelled as 3D element and rebars are modelled as rods. For Concrete, C3D8R element is used and for reinforcing bars HYSD or GFRP, T3D2 element is used from ABAQUS family. The C3D8R element is a linear brick element, with reduced integration. T3D2 element is 2 noded 3 dimensional elements. The longitudinal and transverse reinforcement in the beam is modelled using T3D2. The reinforcement elements in model are assigned with material properties as mentioned in table 1. The reinforcement bars are embedded in the concrete with no slippage or zero slippage for analysis purpose. The boundary conditions are to be assigned at the supports. The load is applied as displacement control in the model. After feeding all the data, ABAQUS runs the prepared model. After completing the analysis, results are recorded in .odb file, and the final results are constructed in the form of plots for our model. The nomenclature for the two cases of beams is mentioned in table 2.

Table 2. Beam Nomenclature.

Nomenclature	Beam description
BLSSS	Beam with Longitudinal rebars as Steel and Stirrups as Steel rebars
BLGSS	Beam with Longitudinal rebars as GFRP and Stirrups as Steel rebars

5. Model verification

The RC beam specimen which was experimentally tested by Tejaswini and Rama Raju [7] has been modelled to verify the correctness of the model. The RC beams has been modelled with the same dimensions and parameters as mentioned in their work. The beam had a cross section of 100 mm x 200 mm with span length of 1200 mm. The grade of concrete is considered as M30 and grade of steel as Fe415. The beam is provided with two numbers of 12 mm diameter rebars as tension steel reinforcement. For the model verification, the FEM results are compared with the experimental results. The variance in results was within 90%, hence correctness of FE model is verified. Figure 2 shows the comparison of load versus deflection plots for RC Beam. It is observed that the numerical results obtained from the

FEM analysis are in good agreement with the experiment results.

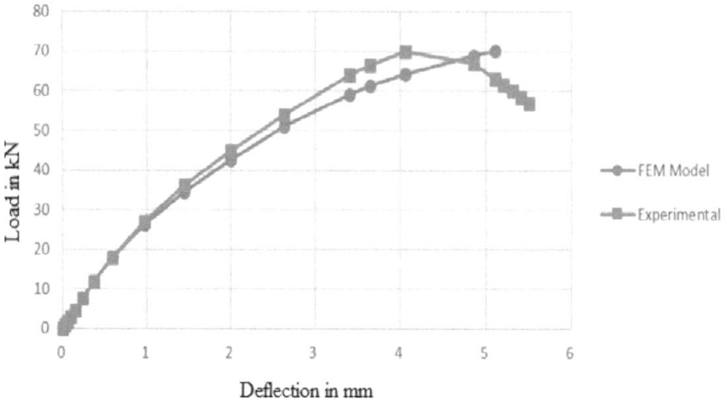

Figure 2. Comparison of load versus deflection plots for RC Beam.

6. Results and discussions

Figure 3 and figure 4 shows the stress distribution in reinforced bars at failure for BLSSS and BLGSS beams respectively. The ultimate stress at mid-point in steel region for tension in BLSSS beam is 352 N/mm^2 and in BLGSS beam is 320 N/mm^2. The gradient in stress distribution is decreasing from mid-point towards the end supports in both the beams.

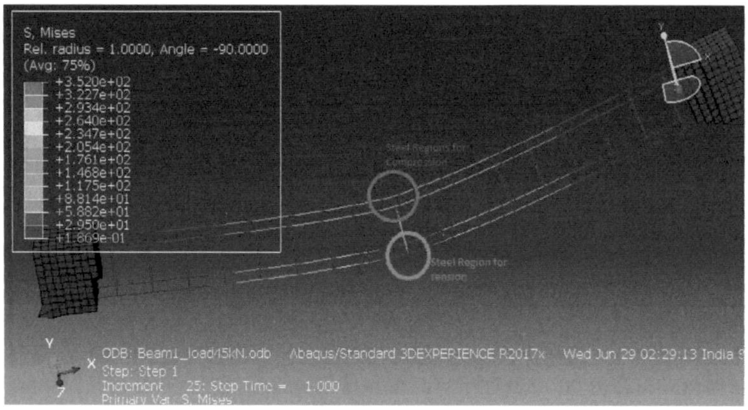

Figure 3. Stress distribution in reinforced bars at failure (BLSSS).

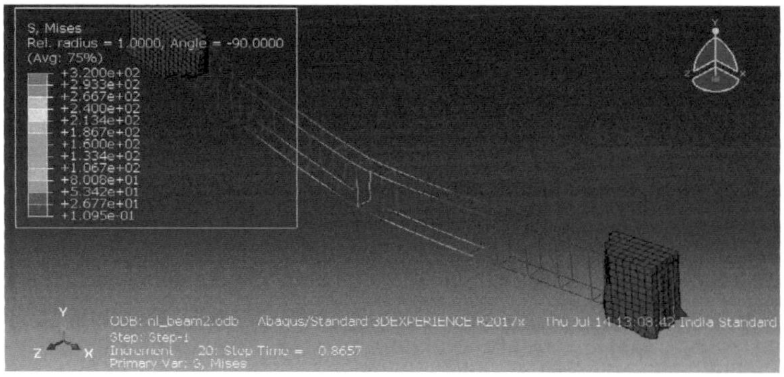

Figure 4. Stress distribution in reinforced bars at failure (BLGSS).

Figure 5 represents stress versus strain variation in top element (concrete) at mid-point of the BLSSS beam and BLGSS beam. From the plot, it can be seen that for beam BLSSS curve is linear till the stress value of 7 N/mm^2. There after non linearity is observed in plot with compressive stress increasing to the peak value of 30 N/mm^2 at a compressive strain of 0.0024. After reaching the peak, strain softening is observed till a strain value of 0.0041. For the BLGSS beam, curve is linear till the stress value of 7 N/mm^2. There after non linearity is observed in plot with compressive stress increasing to the peak value of 30 N/mm^2 at a compressive strain of 0.0048. Compared to the conventional BLSSS beam, the ultimate stress in beam BLGSS has occurred at a higher strain value. From the plots, it can be seen that BLSSS beam has undergone softening in the process of application of load, whereas strain in concrete in BLGSS beam is continuously increasing till the failure of specimen without softening.

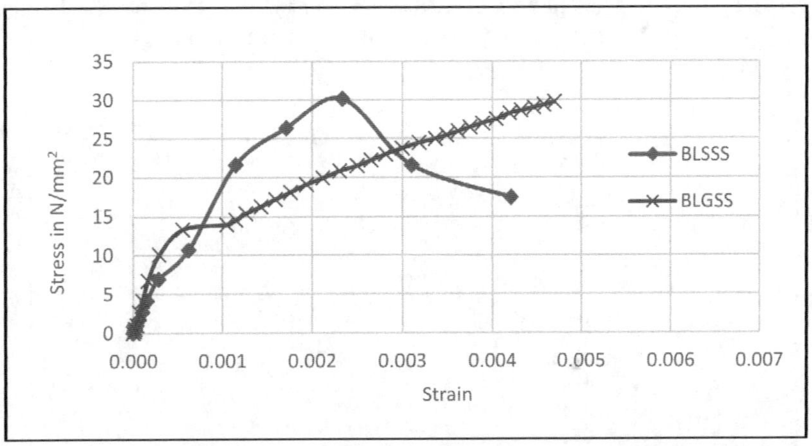

Figure 5. Stress versus strain curves of concrete for the BLSSS & BLGSS beams.

Figure 6 represents load versus strain curves in tension bar for the BLSSS beam and BLGSS beam. From the figures, it can be observed that trend of the plots is bilinear for the BLSSS and BLGSS beams. From the plot, it can be seen that, for BLSSS beam curve is linear till the load of 12 kN, when the strain in rebar is 0.0002. This is the point at which the yielding of reinforcing bar is taking place. Further the yielding of this rebar continues till the failure of beam specimen at a load of 45 kN with the ultimate strain value in rebar of 0.0018. For the BLGSS beam curve is linear till load on the beam is 7 kN. This is the point at which the yielding of reinforcing bar is taking place. Further yielding of this rebar takes place till the failure of beam specimen at a load of 38 kN with the ultimate strain value in rebar of 0.0035. Compared to conventional BLSSS beam, the strain developed in tension bars is of higher value in BLGSS beam which is an indicator of more ductility.

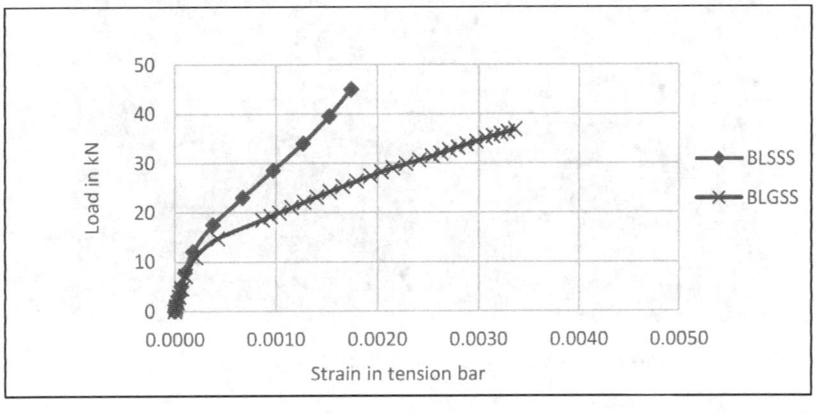

Figure 6. Load versus strain curves in tension bar for the BLSSS & BLGSS beams.

Figure 7 represents load versus deflection curves for the BLSSS beam and BLGSS beam. From the plot, it can be seen that for BLSSS beam, curve is linear till the load of 12kN, with the corresponding displacement of 0.65 mm. After this point, curve is varying in nonlinear manner till failure. From the figure, it can be noted that maximum load carrying capacity of beam is 45 kN and the corresponding deflection is 4.7 mm. For the BLGSS beam curve is linear till the load of 11 kN, with the corresponding displacement of 0.75 mm. After this point, curve is varying in nonlinear trend till failure with a maximum load carrying capacity of 38 kN and the corresponding deflection of 6.8mm. From the results, it is observed that load carrying capacity of BLGSS beam decreases by 16% as compared to the conventional beam. Further displacement of BLGSS beam increased by 45% as compared to conventional beam. The area under the load versus deflection curve, represents the energy absorption in the beam, is calculated and determined as 0.1328 kNm and 0.1713 kNm for the BLSSS beam and BLGSS beam respectively. Further on comparison, BLGSS beam absorbed 29% more energy than conventional beam BLSSS.

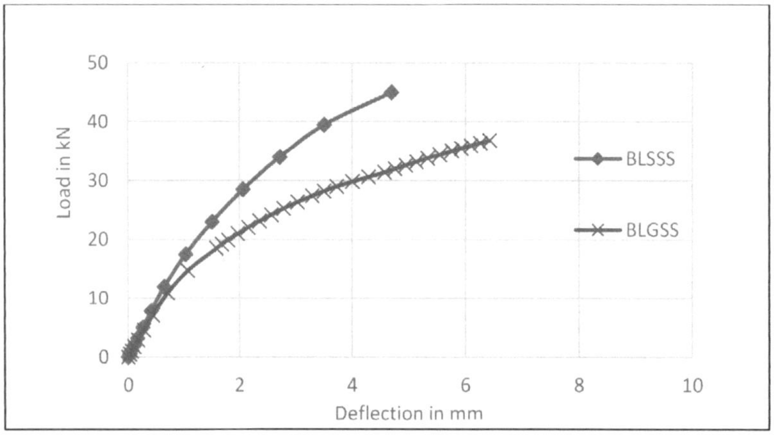

Figure 7. Load versus deflection curves for the BLSSS & BLGSS beams.

Figure 8 and figure 9 shows deflection at failure for the BLSSS beam and BLGSS beam respectively. The deflection at failure at mid-point is 5.877 mm and 6.938 mm in BLSSS beam and BLGSS beam respectively. The deflection profile is decreasing from mid-point towards the end supports in both the beams.

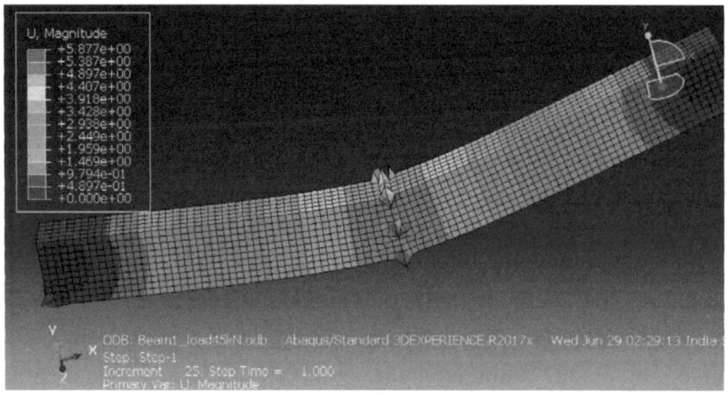

Figure 8. Deflection at failure (BLSSS).

IOP Conf. Series: Earth and Environmental Science 1149 (2023) 012011 doi:10.1088/1755-1315/1149/1/012011

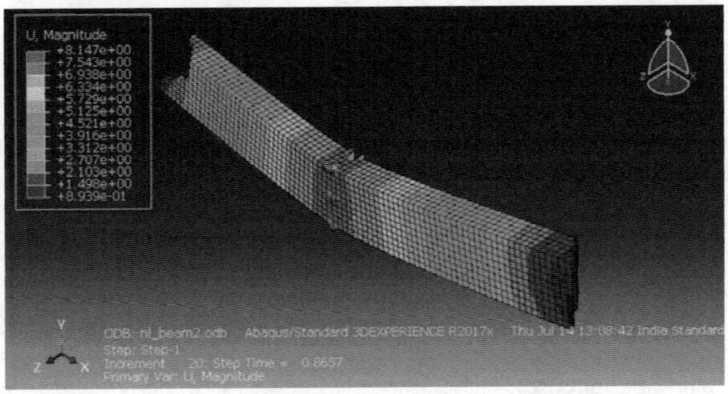

Figure 9. Deflection at failure (BLGSS).

7. Conclusions

From the above numerical investigation, the following conclusions are made:

- It has been observed that BLSSS beam has undergone softening in the process of application of load, whereas strain in concrete in BLGSS beam is continuously increasing till the failure of specimen without softening.
- The strain developed in tension bars is of higher value in BLGSS beam as compared to the conventional BLSSS beam which is an indication of more ductility. On quantitative basis, strain in tension bar of GFRP bars increased by 90% as compared to longitudinal steel rebar in conventional beam.
- The load carrying capacity decreases by 16% when longitudinal steel bars are replaced by GFRP bars.
- Mid-point deflection for the beam with longitudinal GFRP bars increased by 45% when compared to conventional beam BLSSS.
- The energy absorption of BLGSS beam is 29% more than that of conventional beam BLSSS.

References

[1] Salih R and Zhou F Y 2019 Numerical investigation of the behaviour of reinforced concrete beam reinforced with FRP bars *Civil Engineering Journal* **5(11)** pp 2296-308

[2] Panahi M and Izadinia M 2018 A Parametric Study on the Flexural Strengthening of Reinforced Concrete Beams with Near Surface Mounted FRP Bars *Civil Engineering Journal* ***4*(8)** pp 1917-29

[3] Hamad R, Johari M M and Haddad 2019 Effects of bars slippage on the pre-and post-heating flexural behaviour of FRP reinforced concrete beams experimental and theoretical investigations *International Journal of Civil Engineering and Technologies* ***10(2)*** pp 574-602

[4] Ge W J, Ashour Ashraf F, Cao D F, Lu W, Gao P, Yu J, Ji X and Cai C 2019 Experimental study on flexural behaviour of ECC-concrete composite beams reinforced with FRP bars *Composite Structures* ***208(6)*** pp 454-65

[5] Protchenko K and Szmigiera E 2020 Post-fire characteristics of concrete beams reinforced with hybrid FRP bars *Materials* ***13*(5)** p 1248

[6] Hekal, Salama, Elsamak and Almaadawy 2021 Numerical Investigation of Shear Behaviour of RC T-beams Strengthened by Ultra High-Performance Fiber Reinforced Concrete (UHPFRC) *IOSR Journal of Mechanical and Civil Engineering* **18(4)** pp 13-28

[7] Tejaswini T and Rama Raju M V 2015 Analysis of RCC Beams using ABAQUS *International Journal of Innovations in Engineering and Technology (IJIET)* **5(3)** pp 248-55

A Comparative Study on Dynamic Response of Buildings Resting on Coir and Rubber mat Reinforced Soil Bed

Sreya M V [a], B R Jayalekshmi[b] and Katta Venkataramana[c]
[a]Assistant Professor, Department of Civil Engineering, St. Joseph Engineering College, Vamanjoor, Mangalore 575028, India.sreyav@sjec.ac.in (Corresponding author)
[b]Professor, Department of Civil Engineering, National Institute of Technology Karnataka, Surathkal, Mangalore 575025, India. brjaya@nitk.edu.in
[c]Professor, Department of Civil Engineering, National Institute of Technology Karnataka, Surathkal, Mangalore 575025, India. ven.nitk@gmail.com

Abstract. Geotechnical seismic isolation has emerged as an efficient technique for mitigating the severe effects of earthquakes by providing smooth synthetic liners beneath foundations or between soil layers for dissipating seismic energy through sliding. This study investigates the efficacy of using a rubber mat and a natural coir mat as reinforcement materials within the soil to act as a seismic soil-isolation medium. A three-dimensional finite element simulation of five-storey buildings resting on raft foundations in soft soil with and without the soil-isolation mechanism has been performed. The reinforced soil-structure system was exposed to two different earthquake motions, such as the ground motions corresponding to the elastic design spectrum for Zone III as per the Indian standard code (IS 1893 (Part 1): 2016) and the Northridge earthquake (1994). The proposed study deals with the analysis of dynamic responses of buildings when the soil is reinforced with a coir mat and rubber mat under earthquake motions. The findings show that the seismic responses of low-rise buildings are significantly reduced by a novel technique proposed in this work to reinforce the soil with isolation materials in their mat form to reduce the seismic responses under earthquake loads.

Keywords. *Soil isolation-building system, raft foundation, coir mat, rubber mat, finite element modelling*

1. Introduction

Earthquake is one of the major disasters that destroy structures around the world. Buildings need to be built to withstand earthquakes, or their destructive effects need to be mitigated by adopting seismic isolators to provide secure living conditions. When structural isolation is used, the damaging effects of earthquakes on the structure and its building contents can be lessened by creating a sliding or flexible mechanism between the structure and the foundation to separate the lateral movements of the structure from the lateral movements of the soil. However, in soil isolation, soil-reinforced material absorbs seismic energy before it reaches the structure.

A variety of seismic isolation techniques are available from the previous literature in the area of soil isolation. The benefits of using a geosynthetic reinforced isolation system to reduce seismic response during ground vibrations were first studied by Martin, P P; Seed (1982). The provision of rubber sand mix (RSM) surrounding and below the foundation of the structures to act as a cushion [2] and the provision of smooth synthetic liners below the foundation or within the soil at a distance below the footing were explored by researchers [3]. Economical seismic isolation strategies can be useful for nations when there are insufficient resources and/or technology to reduce earthquake severity. Geotechnical isolation refers to seismic isolation techniques that use geotechnical methods, while structural isolation refers to techniques that incorporate structural methods, such as vibrators in superstructures.

Another reinforcement material is coir, which was applied in previous studies in their fibre form in soil to strengthen it and prevent liquefaction [4]. At greater shear strains, where its contribution significantly improves the shear modulus (G), the coir fibre performs better. However, there are some questions about how long the coir material will last. By applying a phenol and bitumen coating on the fibers, the durability can be increased. One method for enhancing the mechanical and durability properties of reinforcing material is chemical treatment [5]. The importance of using natural material as isolation material in their mat form is found to be a novel technique in this scenario. Because isolation materials give better efficiency in soil isolation process when it is provided as mat as compared to fibers.

2. Methodology

Conventional reinforcement materials like natural fibers, rubber tires and geomembranes, strengthen the soil under static loads. Even under the dynamic loads also materials like geomembrane were used in the previous studies. But getting an alternative sustainable material that can act as an isolation material under earthquake loads is a difficult task. Therefore, in this study, an attempt to study the efficacy of materials like rubber and coir in their mat form is used and compared the efficiency of the materials to reduce the seismic responses under the dynamic loads. Due to the large frictional resistance at the interface of soil and foundation the seismic energy transferred from the soil to structure will also be higher when the foundation rests on soil which do not isolated from the earthquake loads.

The seismic response at the top of the structure was simulated in this study using five storey, three-dimensional building for both the isolated and non-isolated states of soil. soft soil was used for the model. The input parameters used for modelling building, soil and structure components are shown in Table 1. A sum total of 4kN/m^3 load was applied on each floor of the building and the self-weight of the structural mass also was considered. Finite element based ANSYS software was used for modeling and studying the soil-structure interaction under dynamic loading. The elements used for modelling the soil-raft-structure components and their details are shown in Table 2.

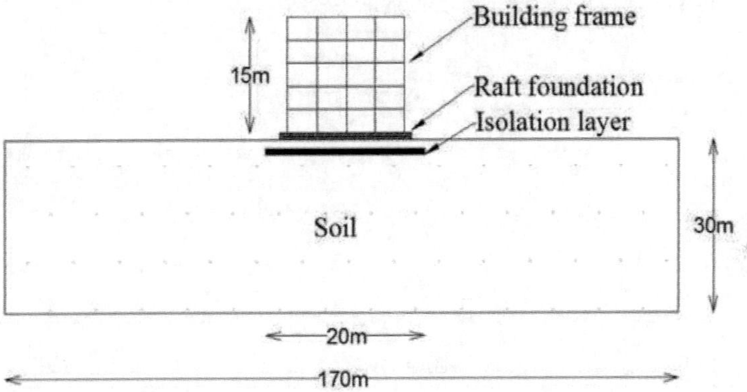

Fig. 1 Schematic illustration of the model

In order to replicate the infinite behavior of soil viscous boundaries were provided on the lateral sides of soil stratum so that no waves will reflect back once it goes from the soil to lateral boundaries. At the bottom of soil both horizontal and vertical translations are prevented and act as a rock surface.

The entire soil-structure system was subjected to an external input motion to study the response under earthquake motion. Input motions used are the accelerogram recorded corresponding to the elastic design spectrum for Zone III as per the Indian standard code (IS 1893 (Part 1): 2016) and the scaled Northridge earthquake (1994) having 0.3g peak ground acceleration. Seismic response of both soil-structure system incorporating both reinforced and unreinforced cases of soil are analyzed. Coir mat and rubber mat were placed at 1m below the foundation in soil, which is the optimum value of placement [6]. Absolute deflection of the roof mass is the main response structural quantity to assess the performance of isolation system, since it is proportional to the bending moment and shear force that exerted at the bottom of the superstructure.

Table 1. Properties of building components and isolation materials

Specification	**Dimension**	**Young's Modulus of Elasticity (MPa)**	**Poisson's ratio**	**Unit weight (γ_d) (kN/m^3)**
RC Beam (BxDxL)	0.3mx0.3mx4m	25x10^3	0.15	25
RC Column (BxDxL)	0.5mx0.5mx3m	25x10^3	0.15	25
Raft foundation (BxDxL)	18mx18mx1m	25x10^3	0.15	25
Rubber mat (BxW)	20mx20m	100	0.49	15
Coir mat (BxW)	20mx20m	4100	0.3	15
Soft soil	170mx170mx30m	65	0.4	16

Table 2. Elements and details used for modelling in ANSYS

Component	**Element**	**Details**
Soil	SOLID 185	8-noded first order linear brick reduced integration element.
Raft foundation		Good for convergence in contact analysis
Building	BEAM188	2-noded beam element with linear behavior
Coir mat/ Rubber mat	SOLSH190	3D finite strain 190 element having negligible bending stiffness

3. Results and discussions

The peak roof deflection responses of building under seismic loading were estimated. Fig. 2 represents the maximum value of roof deflection observed for the isolated conditions of soil, i.e. with coir mat and rubber mat placed at its optimum thickness (1m) and for conventional soil in the present study. Compared to the unreinforced soil bed, the reinforced soil bed shows reduced building responses under seismic loading. Deflection shown by the building resting on soil reinforced with coir mat is less than that observed with rubber mat. This shows that the material, coir mat is better in absorbing the vibration energy before it reaches to the structure from soil.

Base shear ratio (F') for the five storey building was calculated using Eq^n,

F'= F/W, where, w is the weight of the superstructure and F is the total shear force calculated at the ground floor.

The base shear ratio obtained for the models considered for the analysis shows that the soil reinforcement isolation mats such as coir mat and rubber mat considerable reduces the seismic base shear and the coir mat shows much better isolation efficiency over rubber mat. Fig. 3(a) and Fig. 3(b) show the base shear ratio and the percentage reduction in the seismic base shear ratio respectively for different soil-structure system considered.

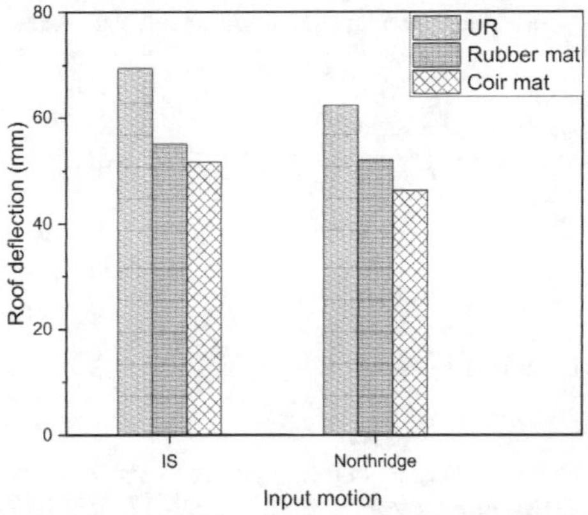

Fig. 2 Roof deflection for soil reinforced with coir mat and rubber mat under different earthquake input motions

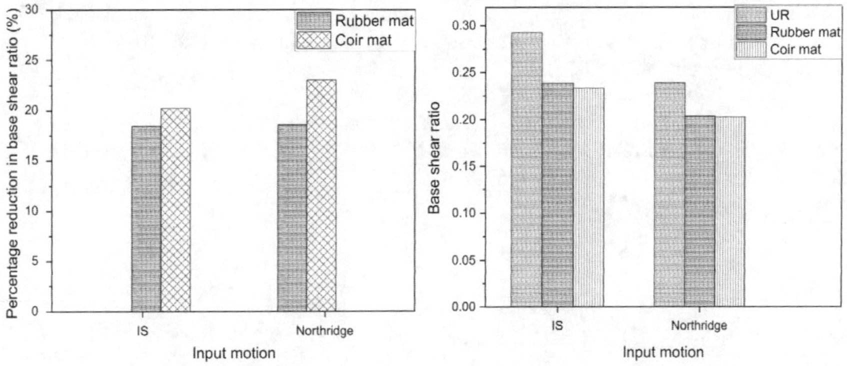

Fig. 3 (a) Seismic base shear ratio of building on reinforced soil with various mats (b) Percentage reduction in the seismic base shear ratio

4. Conclusions

The isolation efficacy of the coir mat and rubber mat in reducing the seismic response of building, those positioned under the foundation, is explored in the current numerical analysis. The following are the conclusions arrived:

1. The soil reinforced with coir mat and rubber mat could considerably reduce the seismic building responses as compared to a conventional building.
2. Maximum attenuation in deflection at roof of buildings occurred by reinforcing soil with coir mat i.e. 28% and at the same time rubber mat could reduce 22% of seismic deflection compared to unreinforced soil case.

3. Seismic base shear is found to be less when soil reinforced with coir mat and it shows 24% reduction in seismic base shear ratio.
4. Materials show same isolation efficiency under both the input motions considered. Which shows that the frequency of input motions which matches with the fundamental frequency of soil-structure system considered was abundant.
5. Seismic isolation with coir/rubber mat was found to be a simple and alternative technique to reduce earthquake severity than the available traditional base-isolation methods.

References

[1] H. B. Martin, P P; Seed. Rock and Soil Improvement Techniques. p. 836288, 1982.

[2] M. K. Yegian and M. Catan. Soil Isolation for Seismic Protection Using a Smooth Synthetic Liner. *J. Geotech. Geoenvironmental Eng.*, vol. 130, no. 11, pp. 1131–1139, 2004, doi: 10.1061/(asce)1090-0241(2004)130:11(1131).

[3] T. H. Ho. Geotechnical seismic isolation. *Earthq. Eng. New Res.*, no. January 2011, pp. 55–87, 2011.

[4] A. Boominathan and S. Hari. Liquefaction strength of fly ash reinforced with randomly distributed fibers. *Soil Dyn. Earthq. Eng.*, vol. 22, no. 9–12, pp. 1027–1033, 2002, doi: 10.1016/S0267-7261(02)00127-6.

[5] A. Kumar and D. Gupta. Behavior of cement-stabilized fiber-reinforced pond ash, rice husk ash-soil mixtures. *Geotext. Geomembranes*, vol. 44, no. 3, pp. 466–474, 2016, doi: 10.1016/j.geotexmem.2015.07.010.

[6] M. V Sreya, B. R. Jayalekshmi, and K. Venkataramana. A study on seismic response of buildings on coir mat reinforced sand bed. *IOP Conf. Ser. Mater. Sci. Eng.*, vol. 1114, no. 1, p. 012018, Mar. 2021, doi: 10.1088/1757-899x/1114/1/012018.

IOP Conf. Series: Earth and Environmental Science 1149 (2023) 012013 doi:10.1088/1755-1315/1149/1/012013

Fillers Influence on Hot-Mix Asphalt Mixture Design and Performance Assessment

Vinay Hosahally Nanjegowda[a], Rathankumar M N[b], and Anirudh N[c]

[a] Assistant Professor, Department of Civil Engineering, Siddaganga Institute of Technology, Tumakuru, Karnataka 572103, India, E-mail: vinayhn@sit.ac.in
[b] Graduate Researcher, Department of Civil Engineering, Siddaganga Institute of Technology, Tumakuru, Karnataka 572103, India.
[c] Assistant Professor, Department of Civil Engineering, Jawaharlal Nehru National College of Engineering, Shimoga, Karnataka 577204, India

Abstract: Filler's presence in hot-mix asphalt (HMA) is even though minimal but they do affect its durability characteristics. Many natural and waste materials in the form of fillers have been studied for their effectiveness on HMA mix design and performance characteristics. However, in practice, stone dust (SD) is the preferred filler due to its abundance, ease of availability, and cost-effectiveness. Thus, the major objective of this study was to investigate the effect of locally available materials: stone dust (SD)- natural, hydrated lime (HL) – processed, rice-husk ash (RHA) and fly-ash (FHA)-waste materials on HMA properties based on the factors such as availability, field utilization, cost, and sustainability, while at the same time identify the anomalies of those selected fillers on HMA mix if any. A viscosity grade (VG-30) binder was selected and checked for its fundamental consistency characteristics set forth in Indian standards. In this study, aggregate gradation structure specified as bituminous concrete grading 1 (BC^1) in India was designed for the preparation and evaluation of four HMA mixes: (a) BC^1 with SD (BC^1-SD), (b) BC^1 with RHA (BC^1-RHA), (c) BC^1 with FA (BC^1-FA), and d) BC^1 with HL (BC^1-HL). Fillers: RHA, FA, and HL were studied for their physico-chemical properties. The most recommended filler dosage of 4% by weight of mix was selected and kept uniform for the various BC^1 mixes. Marshall method of mix design was performed to identify the optimum asphalt content (OAC) of four different BC1 mixes. The test results of methylene blue value (MBV), german filler value (GFV), and fineness modulus (FM) indicated that RHA includes more micron-to-nano sized particles than the other two fillers (HL and FA). Thescanning electron microscope coupled with energy dispersive X-ray results showed that the RHA and FA exhibited similar chemical composition, while HL was identified to be a calcium-based compound. The BC^1-RHA mix resulted in non-cohesive mix for the binder content ranging from 4.5 to 6.5%. Additionally, for the binder contents in the range of 7 to 9% the BC^1-RHA compacted samples failed to yield air voids of 4% required to arrive at the OAC. The BC^1- FA mix showed the highest Marshall stability (26.97 KN) followed by BC^1-HL (23.97 kN), and BC^1-SD (17.9 kN). Also, retained stability test results of all the three differentmixes were in close proximity to each other indicative of the affinity of the fillersto asphalt. The resistance to moisture susceptibility results indicated that HL is thebetter anti-stripping element followed by FA, and SD. Among the three different filler-based BC^1 mixes, BC^1-HL mix was adjudged as an effective moistureresistant mix followed by BC^1-FA, and BC^1-SD. However, a single filler that not only tends to improve the various performance parameters of the mix but be available in abundance and cost-effective is yet to be explored.
Keywords: Fillers; Physico-chemical; Marshall stability; Retained stability; Moisture-susceptibility.

1. INTRODUCTION

The major composition of a conventional hot-mix asphalt (HMA) are aggregates, filler, and binder. It is well known that fillers are those particles that passes the sieve size of 0.075 mm (# 75 micron). Filler's contribution to the properties of HMA mixtures subjected to transient loading and climatic conditions are studied based on types and concentrations [1-7]. Fillers tend to improve the asphalt viscous flow and delay aging process [8]. Many researchers believe that the asphalt mastic created in the asphalt mixture is due to the interaction between asphalt and fillers [9,10]. Antunes and co-workers[11], through chemical analysis and asphalt-filler mastic tests concluded that the critical concentration of filler in asphalt is a function of particles geometrical characteristics than their chemical composition.The filler interaction ability with asphalt is dependent on its concentration, size, silica (SiO_2), and the asphalt components [12]. Some researchers postulated that, few fillers are inert in nature with no significant chemical interaction or low reactivity between the binder and filler [13]. A substantial literature pertaining to the effect of fillers classified under natural, industrial wastes, construction and demolition wastes, mining wastes, domestic / municipal wastes, and agricultural by-products on HMA properties are available in research repository [7, 14-18]. However, the most utilized fillers in HMA arestone dust (SD), cement and hydrated lime (HL) [7, 18-20].

Rice is one of the majorly consumed food products in the world. More than 90% of the world's rice is produced and consumed in Asian continent [21]. During rice milling, the rice husk (RH) is removed and merely dumped as waste material. Globally, approximately 731 million tons (MT) of RH is produced [22] with India's contribution being around 127 MT [23]. Some proportion of the RH generated in India and other Asian countries is used as a fuel for: (a) biomass power plants, (b) brick kilns, (c) cardboard makings, (d) mushroom cultivation, and (e) biomass cookstoves in rural areas [24], while about 60% of RH are burnt in open field as a compulsion rather than choice [25]. Research studies suggest that combustion of RH in open field generates greenhouse gases (GHG) that would include polycyclic aromatic hydrocarbons, polychlorinated dibenzo-p-dioxins, and polychlorinated dibenzofurans [26]. In terms of GHG emissions, about 0.7-4.1 g of methane, 0.019-0.057 g of nitrous oxide, carbon dioxide, sulphur dioxide, and volatile organic components are produced due to burning of RH [27]. In addition to GHG emissions RH burning also results in generation of particulate matter (PM) of sizes PM10 and PM2.5 that affects the air quality and visibility [28]. A compositional analysis of rice-husk ash (RHA) indicated SiO2 to be the predominant compound followed by aluminium oxide, while iron oxide, calcium oxide, and sodium oxide as trace elements [29]. Few research studies on the usage of RHA in asphalt concrete as an alternative filler to conventional materials have been established [30-34]. Marshall mix design method was conducted by replacing limestone with RHA in proportions of 25, 50, 75, and 100% [33]. The Marshall stability was found out to be highest for the mix with 50% RHA and 50% limestone. Another comparative study between RHA and cement-based surface course mixtures for Marshall stability, water sensitivity, and long-term aging was conducted [34]. The test results of RHA mixture showed increased strength and durability than that of cement treated asphalt mixture, but the percentage air voids (Pa) increased for RHA than cement.

Most pavement agencies define the selection of fillers on their availability, field utilization, and cost. Thus, the major objective of this study was to investigate the effect of locally available fillers: SD (quarry dust), HL (processed), RHA and fly-ash (FHA) (waste materials) fillers on HMA properties by considering their availability, field utilization, cost, and sustainability factors, while at the same time identify the anomalies of those selected fillers on HMA mix if any.

2. EXPERIMENTAL METHODOLOGY

2.1 Materials

Stone dust (SD) processed at crusher plant was selected as conventional filler. The RHA was collected from a single rice mill factory located in Tumakuru (13° 28' 3.684"N 77° 2' 22.38" E). The cost of RHA was about 12 rupees per kg. FA as filler was collected from a steel industry located in Bellary (15° 12' 33.012"N 76° 39' 59.94"E), at a base price of 32 rupees per kg. HL was another filler that was obtained at a market price of 400 rupees per kg. The fillers were included in the HMA mixture as procured. The aggregates of different sizes were collected from a crushing plant in Tumakuru (13° 28' 28.524"N 77° 2' 1.932"E). Additionally, the aggregates of different sizes were sampled to prepare laboratory designed asphalt concrete samples. The asphalt binder of viscosity grade (VG30) was procured from Mangaluru (13° 4' 20.064"N 74° 47' 54.672"E). Geographically, all the places are in India.

2.2 Material Characterization Methods

2.2.1. Fillers:

The particle size distribution, relative density and plastic limit (PL) properties of RHA, HL and FA was identified according to [35]. Methylene blue value (MBV) test was conducted to identify the presence of harmful clays, organic matter, and any other impurities in fillers as per [36]. German filler test as reported by [9] was conducted to determine the porosity and cohesive nature of the fillers. The hydrophilic nature of the selected fillers was identified using hydrophilic co-efficient test according to [37]. Further advanced material characterization technique scanning electron microscope (SEM) was used. SEM model (JEOL JCM 6000Plus) coupled with energy dispersive X-ray (EDX) analyzer was employed to obtain the surface morphology and the elemental composition of the RHA, FA, and HL.

2.2.2. Asphalt Binder:

VG30 binder was used in the preparation of different HMA mixtures. Fundamental consistency tests such as specific gravity, penetration, ring and ball, and ductility tests were conducted as per [38]. Table 1 provides the fundamental consistency test data of VG30 binder that was compared with the specification values [39].

Table 1. Conventional Binder Consistency Test Results.

Tests	Results	Specified limit
Penetration at 25 °C, mm	66	Min 45
Ductility at 25 °C, cm	75	Min 40
Softening point, °C	48	Min 47
Specific gravity	1.01	0.97 – 1.02
Absolute viscosity at 60 °C, cP	2600	Min 2400
Kinematic viscosity at 135 °C, cP	1425	Min 350

2.2.3. Aggregates:

In this study, three sizes of designated aggregates were used to produce different filler-based asphalt mixtures. Multiple aggregate stockpiles of 20 mm passing all-in-aggregates, 10 mm passing all-in-aggregates, and 4.75 mm passing all-in-aggregates were employed. Aggregates of various sizes were tested using protocols of [40-42] as recommended by [43]. The physical test results of the aggregates are depicted in Table 2. In this study aggregate gradation structure as shown in Table 3 was formulated for bituminous concrete grading 1 (herein referred to as BC^1) specified in [43]. The various aggregate sizes and their proportions for BC^1 was achieved by combining the aggregates from multiple stockpiles by trial-and-error method (Table 3).

Table 2. Aggregates Physical Properties Test Results.

Tests	Results	Specified limit
Aggregate impact value	25.09%	Max 27%
Aggregate crushing value	26.86%	Max 30%
Loss angeles abrasion value	28.12%	Max 30%
Water absorption value	0.62%	Max 2%
Specific gravity		
(20 – 10 mm)	2.679	-
(10 – 6.3 mm)	2.621	-
(4.75 – 0.075 mm)	2.640	-
Combined flakiness and elongation index	26.15%	Max 35%
Angularity number	9.00	0 - 11

Table 3. Developed Aggregate Gradation for BC[1].

I.S Sieve mm	Cumulative percentage by weight of total aggregates passing	
	MoRT&H limits	Achieved aggregate gradation
26.5	100	100
19	90-100	95
13.2	59-79	71
9.5	52-72	64
4.75	35-55	43
2.36	28-44	34
1.18	20-34	27
0.6	15-27	21
0.3	10-20	13
0.15	5-13	7
0.075	2-8	4

2.2.4. Asphalt Mixtures:

A filler dosage of 4% was kept uniform for all the BC[1] mixes. The laboratory mixing and compaction temperatures for the BC[1] mix with VG30 binder were identified to be 165 °C and 151 °C respectively according to equi-viscous method [44]. The mixing and compaction temperatures were derived in accordance with the standard procedure [45] with the help of Equation (1 and 2). The cylindrical samples were compacted with 75 blows on each end of specimen to simulate heavy traffic condition. Marshall mix design procedure enumerated in [44] was adopted to arrive at the various volumetric properties of the BC[1] mixes. Overall, four different BC[1] mixtures: a) BC[1] with SD (BC[1]-SD), b) BC[1] with RHA (BC[1]-RHA), c) BC[1] with FA (BC[1]-FA), and d) BC[1] with HL (BC[1]-HL). In the present study, BC[1]-SDwas considered as conventional mix due to its widespread application. BC[1]-SD, BC[1]-FA, and BC[1]-HL mixtures were compacted for trial binder content ranging from 4 to 6.5% with 0.5% increment in dosage of asphalt by weight of mix. All the mixtures were prepared and tested for their Marshall stability (MS) which indicates resistance to deformation(strength), retained stability (RS) as per [46], and indirect tensile strength (ITS) test according to [47]. Both RS and tensile strength ratio (TSR) were indicative of moisture sensitivity of the mixes. Fig.1 indicates the Marshall compactor and the Marshall stabilometer used to record the mechanical strength of the HMA mixtures. Fig. 2 depicts the sample conditioning and test procedure adopted to measure the moisture sensitivity of the different fillers based HMA mixtures. Additionally, Fig. 3 shows the Marshall design procedure outline employed for the different filler based HMA mixtures.

$$\log(log(\eta)) = Ai + VTSi \times logTR \quad (1)$$

$$\log(log(\eta)) = 8.5632 - 2.822 \times logTR \quad (2)$$

Where,

η = Kinematic viscosity, cP
A_i = Intercept of the curve
VTS_i = Slope of the curve
T_R = Temperature, degree Rankine

Fig. 1. HMA samples conditioning and test procedure images (a to c): (a) Marshall compactor; (b) compacted samples; (c) Marshall stabilometer.

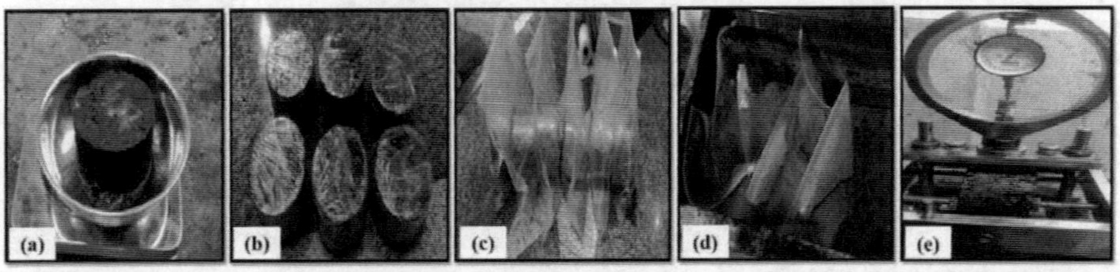

Fig. 2. HMA samples conditioning and test procedure images (a to c): (a) compacted sample; (b) Shrink wrapped samples; (c) Samples placed in vacuum tight covers; (d) typical thawing conditioning; (e) ITS measurement.

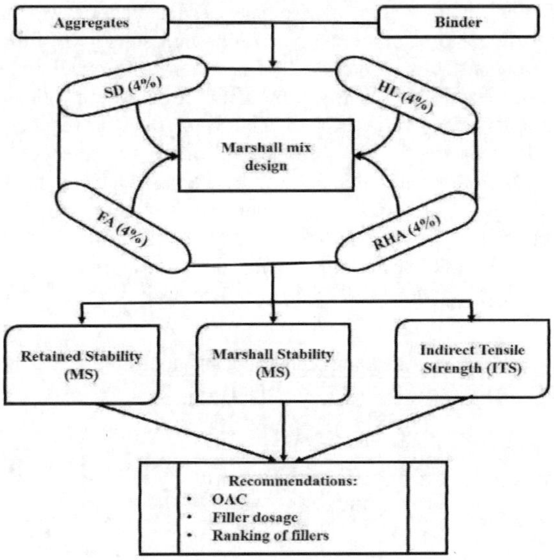

Fig. 3. Flow chart of Marshall mix design procedure with different tests adopted in the study.

3. RESULTS AND DISCUSSION

3.1 Filler Properties

The results of filler characterizations methods adopted in the study are detailed in Table 4. Specific gravity (SG) of FA was found to be slightly higher than that of HL and RHA indicating the particles of same size to be denser than the other two. Methylene blue value (MBV) of RHA (0.50 g/kg) was found to be higher than that of FA (0.43 g/kg) and HL (0.23 g/kg) indicating presence of more clay content. German filler value (GFV) of RHA, HL and FA were 55, 75, and 80 g respectively. Also, the fineness modulus (FM) and D60 values of RHA were lower in comparison to HL and FA (Table 4). The test results of MBV, GFV, FM and D60 indicated that RHA includes more fine sized particles than the other two fillers (HL and FA) and hence more specific surface area. Thus, the absorption capacity of RHA in a liquid medium is higher which would create an in-effective asphalt mixture due to lower binder coating on coarse and fine aggregate particles for a smaller binder dosage. Similar observations were recorded in [9]. Additionally, lower GFV and higher MBV of RHA symbolizes that proportion (mass) ofRHA would affect the asphalt binder and stiffen the HMA mix due to its micron to nano-sized particles.The hydrophilic coefficient values of RHA, HL, and FA were found out to be 0.85, 0.70, and 0.75 respectively. It is indicated that a filler having hydrophilic co- efficient value in the range of 0.75 to 0.85 will have affinity towards bitumen [48].

Table 4. Test Results of Different Fillers.

	Fillers characterizations						
Fillers	**Specific gravity**	**Methylene blue value (g/kg)**	**German filler value (g)**	**Fineness modulus**	**D60 (micron)**	**Hydrophilic coefficient**	**Plasticity index**
HL	2.218	0.23	75	4.96	26	0.70	NP
FA	2.343	0.43	80	6.31	45	0.75	NP
RHA	2.059	0.50	55	3.43	19	0.85	NP

Fig. 4 depicts the micrograph of FA, HL and RHA. The FA particles surface exhibited rough and porous texture, while the pores of FA particles tend to hold rounded molecules (Fig. 4(a)). The SEM image of HL shows the particles to be fluffy, non-porous texture with dissimilarities in dimensions and favour agglomeration (Fig. 4(b)). The SEM image of RHA indicates fibril structure and asymmetrical shape and shows signs of rough texture (Fig. 4 (b)). The EDX technique was used to identify the elemental composition at a particular area of the sample in conjunction with SEM for FA, HL, and RHA fillers. Table 5 presents the concentration of the identified elements in FA, HL, and RHA. In the case of FA, the element Carbon (C) was found out to be the major composition followed by Oxygen (O), Silicon (Si), and aluminium (Al) as shown in Table 5. The Si to Al, Al to O, and Si to O weight (%) ratios were approximately 1:3,1:6, and 1:2 respectively indicative of presence of silica (SiO_2) and alumina (Al_2O_3) pozzolanic material. From Table 5, it can be clearly identified that the hydrated lime is a calcium-based stabilizer having calcium (Ca) as the major concentration (76.22%). The presence of carbonate (CO_3) as an impurity in HL was validated through the identification of C and O as trace elements (Table 5). The filler RHA showed similar chemical composition as FA (Table 5). However, the presence of SiO_2 is more significant than that of Al_2O_3, since Al (1.33) was identified to be the trace element in RHA.

Table 5. Major Elemental Composition of Fillers.

	Elements				
Fillers	**Weight(%)** **C**	**Weigh(%)** **O**	**Weight(%)** **Si**	**Weight(%)** **Ca**	**Weight(%)** **Al**
FA	55.20	26.26	11.09	0.59	4.31
HL	2.53	18.43	0.09	76.22	1.23
RHA	45.56	34.59	12.98	0.0	1.33

Fig. 4. SEM images (a to c): (a) FA-10 kV, 10μ,750×; (b) HL-10 kV, 10μ, 1700×; (c) RHA- 15 kV, 20μ, 700×.

3.2 Mixture Properties

3.2.1. BC[1]-RHA Anomalies:

The Marshall method of mix design adopted for BC1-RHA with 4% RHA (48 g) dosage did not result in compacted samples as shown in Fig 5 for the asphalt content ranging from 4.5 to 6.5% by weight of mix. However, for the binder contents in the range of 7 to 9% by weight of mix with Similar RHA dosage (48 g) resulted in compacted BC[1]-RHA samples but failed to yield air voids of 4% (Table 6) required to arrive at the OAC for BC1-RHA mix. Thus, the anomalies with the selected RHA filler in this study are: (a) the BC[1]-RHA mix showed lower workability (absence of free bitumen) for the asphalt dosages between 4.5 to 6.5%, (b) RHA presence resulted in non-cohesive BC[1]-RHA mix in the asphalt content range of 4.5 to 6.5%, (c) BC[1]-RHA mix failed to yield OAC, while various literatures [15,32,33, 49] for similar aggregate gradation recorded the OAC to be in the range of 4.8 to 5.7% for 5 to 6% RHA dosage, (d) the volumetric properties such as voids in mineral aggregates (VMA), voids filled with bitumen (VFB), percentage air voids (P_a) of BC[1]-RHA compacted samples between binder content range of 7 to 9% did not comply within the specification limits [44] for the production of durable densegraded asphalt mix. The identified anomalies due to RHA filler could be attributed to the physico- chemical properties as enumerated in Section 3.1.

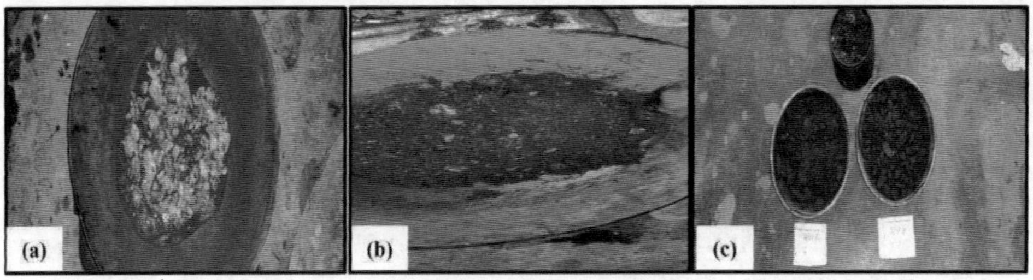

Fig. 5. BC[1]-RHA uncompacted specimens (a) 4.5% binder dosage; (b) 5.5% binder dosage; and (c) 6.5% binder dosage.

Table 6. Volumetric Properties of BC[1]-RHA.

Filler type	Asphalt content (%)	P_a (%)	Bulk density Gmb(g/cm³)	VMA (%)	VFB (%)	MS (kN)	Marshall flow (mm)
	7.0	2.770	2.312	17.745	84.389	11.806	3.100
	7.5	2.046	2.315	18.034	88.653	12.458	3.367
RHA	8.0	1.781	2.306	18.697	90.474	18.894	3.833
	8.5	1.519	2.299	19.348	92.149	16.970	3.967
	9.0	0.827	2.301	19.637	95.789	13.825	5.400

3.2.2. BC[1]-HL, BC[1]-FA, BC[1]-SD Mixtures:

Fig. 6 depicts the various volumetric properties of the three different fillers based HMA mixtures at varying binder contents ranging from 4.0 % to 6.5%. The various mixture volumetric properties were analyzed to identify the OAC at 4% air voids for the HL, FA, and SD based HMA mixtures. Marshall stability (MS) of bituminous mixture is indicative of the maximum load experienced by the sample before failure [33]. Marshall flow value is an indicator for plasticity and flexibility of the mixture under different loading conditions [15,30]. The OAC of BC[1]-HL, BC[1]-FA, BC[1]-SD was found out to be 5.25%, 5.30%, 5.52% respectively from Fig. 6. Among the three mixes, BC[1]-SD conventional mixture required highest asphalt content, while BC[1]-HL, and BC[1]-FA OAC were in close proximity.

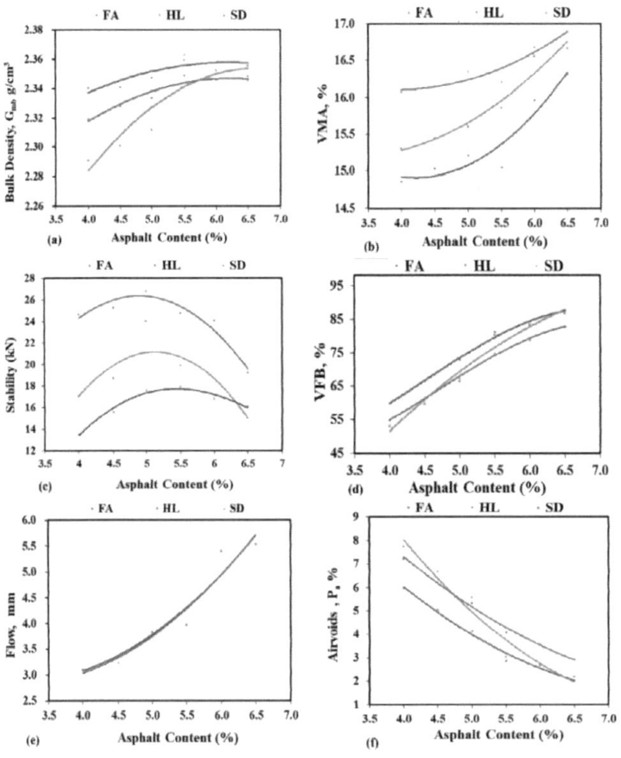

Fig. 6. BC[1]-SD, BC[1]-FA, BC[1]-HL mixtures volumetric properties (a) asphalt content vs Gmb; (b) asphalt content vs VMA; (c) asphalt content vs stability; (d) asphalt content vs VFB; (e) asphalt content vs flow; (f) asphalt content vs air voids.

Table 7 exhibits the mixture volumetric properties of the three mixes at their respective OAC. The BC[1]-FA mix showed the highest MS (26.97 KN) followed by BC[1]-HL (23.97 kN), and BC[1]-SD (17.9 kN). The increased strength characteristics of BC[1]-FA mix could be attributed to the FA chemical composition. The RS test results of all the three different mixes were in close proximity to each other indicative of the affinity of the fillers to asphalt. However, the BC[1]-SD mix had the lowest RS value at 4% filler dosage. Thus, BC[1]-SD mix would provide lower performance for loading and climatic conditions.

Table 7. Volumetric Properties of BC[1]-SD, BC[1]-FA, BC[1]-HL Mixtures.

Properties	Filler types			
	HL	**FA**	**SD**	**MoRT&H**
Filler content (%)	4	4	4	2-8
OAC (%)	5.25	5.30	5.52	Min 5.2
Gmb, (g/cm^3)	2.35	2.37	2.35	-
Marshall stability (kN)	23.97	26.97	17.9	Min 9
Retained stability (%)	88	89	86	Min 80
Pa (%)	3.5	3.8	4.2	3-5
VMA (%)	12.96	15.21	16.45	Min 12
VFB (%)	74	73	75	65-75

3.2.3. Moisture Sensitivity Assessment:

The moisture susceptibility of BC[1]-HL, BC[1]-SD, BC[1]-FA, mixtures were evaluated by investigating indirect tensile strength (ITS) at both unconditioned and conditioned states that helped to determine the tensile strength ratio (TSR). Furthermore, the ITS test at both the conditions for the three different mixtures was carried out for filler dosages ranging from 2 to 5% at an increment of 1% by weight of mix. Fig. 7 represents the ITS test results of the BC[1]-HL, BC[1]-SD, BC[1]-FA mixtures.As observed in Fig. 7(a), BC[1]-HL mix under dry condition showed improved ITS of 25% over 2 to 4% filler dosages. The BC[1]-HL mix at 4% HL dosage showed the highest ITS (1630 kPa) under the dry condition as compared to BC[1]-SD, and BC[1]-FA mixes. However, the ITS reduced slightly at the5% HL dosage (1569 kPa). The ITS of BC[1]-FA mix increased with the increase in the FA dosage (2-5%). The BC[1]-FA mix under dry condition showed improved ITS of 38% from 2 to 4% filler dosages.The BC[1]-FA mix at 5% FA dosage showed similar ITS (1629 kPa) under the dry condition as compared to BC[1]-HL mix. The BC[1]-SD mix at 5% SD dosage resulted in reduced ITS value of 1509kPa under the dry condition. The BC[1]-SD mixes exhibited lower ITS when compared to BC[1]-HL, andBC[1]-FA mixes over the entire specified filler dosages (2-5%) for unsaturated condition (Fig. 7(a)). As observed in Fig. 7(b), BC[1]-HL mix subjected to freezing and thawing temperatures under saturatedconditions showed similar trend as that in dry condition. The BC[1]-HL mixes under conditioned stateshowed improved ITS of 35% over 2 to 4% HL dosages. The BC[1]-HL mix at 4% HL dosage resultedin an ITS of 1430 kPa under the wet condition. As observed in Fig. 7(b), the BC[1]-FA mixes under conditioned state showed improved ITS of 45% over 2 to 4% FA dosages. The BC[1]-FA mix at 4% HL dosage showed the highest ITS (1442 kPa) under the wet condition as compared to BC[1]-HL, andBC[1]-SD mixes. The BC[1]-SD mixes under conditioned state showed improved ITS of 28% over 2 to 4% SD dosages. But, the BC[1]-SD mixes showed reduced ITS relative to BC[1]-HL, and BC[1]-FA mixesover the entire filler dosages as depicted in Fig. 7 (b). Fig. 8 indicates the TSR values of the BC[1]-HL,BC[1]-FA, BC[1]-SD mixtures for the selected filler dosages.

The TSR value of BC[1]-HL mix at 2% of HL dosage was 81% and increased significantly to 87% at 4% HL dosage.Overall, there was an increase of about 7%. Additionally, the BC[1]-HL mix showed the highest TSR value at 4% filler dosage as compared to other two mixes (BC[1]-FA, and BC[1]-SD). Furthermore, the BC[1]-HL mix performed better thanother two mixes over the entire filler dosage range of 2-5% as observed in Fig. 8, indicating that HL used as a filler material in the mix would function asan anti-stripping element. The TSR value of BC[1]-SD mix at 2% of SD dosage was about 81% and increased to 83% at 4% SD dosage. Overall, there was increase of about 2%. The SD filler did not significantly improve the TSR values across the entire filler dosage ranges of 2-5% as observed in Fig.8.

Thus, mere increase in the SD proportion in the BC[1] mix may not simply result in a moisture-resistant mix. The TSR value of BC[1]-FA at 2% of FA dosage was about 78.3% and increased to 82% at 4% FA dosage. Overall, there was an increase of about 4%. The BC[1]-FA showed the lowest TSR values among all the three mixes for the selected filler dosage range of 2-5% as shown in Fig. 8. Thus, FA as a filler may not be prove to be an effective anti-stripping compound. Similar TSR value for HL based HMA mixture was observed in another study [30], that was attributed to the presence of Ca^{2+} ions in HL which acts as a catalyst in creating a chemical bond with aggregate and bitumen making the mixture to arrest moisture from external media. The reduced durability characteristics of FA based HMA mixtures was assigned to the presence of SiO_2 content as the major composition in FA [16,18]. Overall, from the test results of TSR it was adjudged that HL is the better anti-stripping element in presence of water followed by FA, and SD. Among the three different filler-based BC[1] mixes, BC[1]-HL mix was adjudged as an effective moisture resistant mix followed by BC[1]-FA, and BC[1]-SD. This study validated that 4% of HL, FA and SD content by weight of mix would be the optimum filler dosage in a dense graded aggregate structure for strength and durability characteristics.

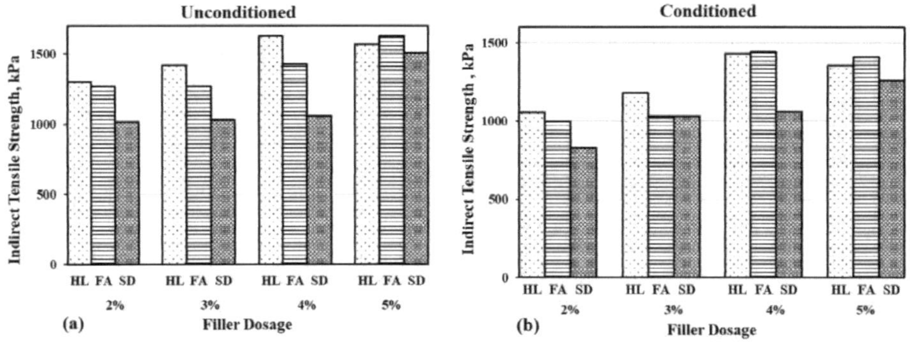

Fig. 7. ITS test results of three different mixtures: (a) unconditioned state; (b) conditioned state.

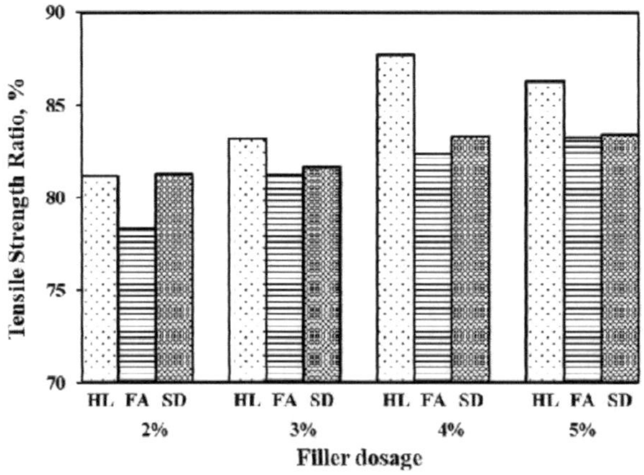

Fig. 8. TSR values of HL, FA, SD fillers based BC[1] mixes.

4. CONCLUSIONS AND FUTURE SCOPE

The major objective of this study was to identify the effectiveness of fillers SD (quarry dust), HL (processed), RHA and fly-ash (FA) (waste materials) on HMA properties based on the factors such as: availability, field utilization, cost, and sustainability. Four types of HMA mixes: (a) BC^1 with SD (BC^1-SD), (b) BC^1 with RHA (BC^1-RHA), (c) BC^1 with FA (BC^1-FA), and d) BC^1 with HL (BC^1-HL) were prepared and assessed for their performance, while drawing the following conclusions from the study:

- The test results of MBV, GFV, FM and D60 indicated that RHA includes more fine sized particles than the other two fillers (HL and FA) and hence for a given asphalt dosage the absorption capacity of RHA is higher which would create an in-effective asphalt mixture due to lower binder coating on coarse and fine aggregate particles. Further, the concentration of RHA in BC^1 aggregate structure plays a significant role in developing a workable BC^1-RHA mixture.
- The Marshall method of mix design for BC^1-RHA with 4% RHA (48 g) dosage did not result in compacted samples for the asphalt content ranging from 4.5 to 6.5% by weight of mix. Additionally, for the binder contents in the range of 7 to 9% the BC^1-RHA compacted samples failed to yield air voids of 4% required to arrive at the OAC.
- The OAC of BC^1-HL, BC^1-FA, BC^1-SD was found out to be 5.25%, 5.30%, 5.52% respectively. Among the three mixes, BC^1-SD conventional mixture resulted in highest asphalt content, while BC^1-HL, and BC^1-FA OAC were in close proximity.
- The BC^1-FA mix showed the highest MS (26.97 KN) followed by BC^1-HL (23.97 kN), and BC^1-SD (17.9 kN). Also, RS test results of all the three different mixes were in close proximity to each other indicative of the affinity of the fillers to asphalt.
- The resistance to moisture susceptibility results indicated that HL is the better anti-stripping element followed by FA, and SD. Among the three different filler-based BC^1 mixes, BC^1-HL mix was adjudged as an effective moisture resistant mix followed by BC^1-FA, and BC^1-SD.

Future scope: In the present study, the selected RHA failed to result in a HMA mixture. However, a comprehensive study for the RHA as a filler on the factors such as: (a) source variability, (b) dosage, (c) different aggregate gradation, (d) binder type, (e) mix design need to be investigated. Thus, a single filler that not only tends to improve the various performance parameters of the mix but be available in abundance and cost-effective is yet to be explored.

ACKNOWLEDGMENTS

The authors gratefully acknowledge the Department of Civil Engineering laboratory facilities at the Siddaganga Institute of Technology Tumkauru to conduct experiments. Further, the authors are thankful to the Centre of Applied Research and Nanotechnology, SIT, Tumakuru and KCTU, Government of Karnataka that helped conduct SEM-EDX experimentation.

DISCLOSURE STATEMENT

The authors declare that there is no conflict of interest.

REFERENCES:

[1] Taha R, Al-Rawas A, Al-Harthy A, Qatan A. Use of cement bypass dust as filler in asphalt concrete mixtures, J. Mater. Civ. Eng. 2002; 14:338-43. https://doi.org/10.1061/(ASCE)0899-1561(2002)14:4(338).

[2] Chandra S, Choudhary R. Performance characteristics of bituminous concrete with industrial wastes as filler, J. of Mater. Civ. Eng. 2013; https://doi.org/0.1061/(ASCE)MT.1943-5533.0000730.

[3] Modarres A, Rahmanzadeh M, Application of coal waste powder as filler in hot mix asphalt, Con. Build. Mat. 2014; 66:476-83. https://dx.doi.org/10.1016/j.conbuildmat.2014.06.002.

[4] Zulkati A, Diew WY, Delai DS. Effects of fillers on properties of asphalt-concrete mixture, J. Transport. Eng. 2012; 138:902–10. https://doi.org/10.1061/(ASCE)TE.1943-5436.0000395.

[5] Zhang H, Li H, Abdelhady A, Mo D, Yang B, Harvey J. Utilization of fine solid waste in asphalt mortar, J. Mater. Civ. Eng. 2019; 31(8):1-12. https://doi.org/0.1061/(ASCE)MT.1943-5533.0002792.

[6] Lima MSS, Thives, LP. Evaluation of red mud as filler in brazilian dense graded asphalt mixtures, Con. Build. Mat. 2020; 260:119894. https://doi.org/10.1016/j.conbuildmat.2020.119894
[7] Choudhary J, Kumar B, Gupta A. Potential utilization of construction wastes in asphalt pavements as fillers using ranking framework, R. Mater. Pave. Des. 2021; 22:624-40 https://doi.org/10.1080/14680629.2021.190569.
[8] Bianchetto H, Miro R, Jimenez FP, Martinez AH. Effect of calcareous fillers on bituminous mix aging, Transportation Research Record, 2007; 1998(1):140-48. https://doi.org/10.3141/1998-17.
[9] Kandhal PS, Lynn CY, Parker F. Characterization tests for mineral fillers related to performance of asphalt paving mixtures, Transportation Research Record 1998;1638(1):101-10.
[10] Kim YR, Little DN, Song I. Effect of mineral fillers on fatigue resistance and fundamental material characteristics: mechanistic evaluation, Transportation Research Record 2003;1832(1):1-8.
[11] Antunes V, Freire AC, Quaresma L, Micaelo R. Effect of the chemical composition of fillers in the filler–bitumen interaction, Con. Build. Mat. 2016; 104:85-91. https://dx.doi.org/10.1016/j.conbuildmat.2015.12.042.
[12] Zhang J, Li X, Liu G, Pei J. Effects of material characteristics on asphalt and filler interaction ability, International Journal of Pavement Engineering 2017; 20(8):928-37. https://doi.org/ 10.1080/10298436.2017.1366765.
[13] Varveri A, Jing R, Tarsi G, Erkens S. Effect of filler properties on the hydrothermal ageing of bituminous mastics, *R. Mater. Pave. Des.* 2021; 22:2-22. https://doi.org/0.1080/14680629.2021.1903975.
[14] Akbulut H, Gurer C, Çetin S, Elmaci A, Investigation of using granite sludge as filler in bituminous hot mixtures, *Con. Build. Mat.* 2012; 36:430-36. https://dx.doi.org/10.1016/j.conbuildmat.2012.04.069.
[15] Arabani M, Tahami SA, Taghipoor M. Laboratory investigation of hot mix asphalt containing waste materials, *R.Mater. Pave. Des.* 2016; 18(3):713-29. https://doi.org/10.1080/14680629.2016.1189349.
[16] Choudhary J, Kumar B, Gupta A. Application of waste materials as fillers in bituminous mixes, *Con. Build. Mat.*2018; 78:417-25. https://doi.org/10.1016/j.wasman.2018.06.009.
[17] Bocci E. Use of ladle furnace slag as filler in hot asphalt mixtures, *Con. Build. Mat.* 2018; 161:156-164. https://doi.org/10.1016/j.conbuildmat.2017.11.120.
[18] Choudhary J, Kumar B, Gupta A. Analysis and Comparison of asphalt mixes containing waste fillers using a novel ranking methodology, *J. Mater. Civ. Eng.* 2020; 32(5): 338-43. https://doi.org/10.1061/(ASCE)MT.1943-5533.0003137.
[19] Antunes V, Freire AC, Quaresma L, Micaelo, R. Evaluation of waste materials as alternative sources of filler in asphalt mixtures, Materials and Structures 2017; 50:254. https://doi.org/10.1617/s11527-017-1126-3.
[20] Valentin J, Trejbal J, Nezerka V, Valentov T, Faltus M. Characterization of quarry dusts and industrial by-products as potential substitutes for traditional fillers and their impact on water susceptibility of asphalt concrete. Con. Build. Mat. 2021; 301:124294. https://doi.org/10.1016/j.conbuildmat.2021.124294.
[21] Kush, Gurudev S. Origin, dispersal, cultivation and variation of rice. Plant molecular biology 1997; 35 (1): 25-34.
[22] Karimi K, Kheradmandinia S, Taherzadeh MJ. Conversion of rice straw to sugars by dilute acid hydrolysis. Biomass and Bioenergy 2006; 30: 247-253.
[23] Karam DS, Nagabovanalli P, Rajoo KS, Ishak CF, Abdu A, Rosli Z, Muharam FM, Zulperi D. An overview on the preparation of rice husk biochar, factors affecting its properties, and its agriculture application. Journal of the Saudi Society of Agricultural Sciences 2022; 21: 149-159.
[24] Gadde B, Bonnet S, Menke C, Garivait S. Air pollutant emissions from rice straw open field burning in India, Thailand and the Philippines. Environmental Pollution 2009; 157: 1554-1558.
[25] Bhattacharyya P, Bhaduri D, Adak T, Munda S, Satapathy BS, Dash, PK, Padhy SR, Pattanayak A, Routray S, Chakraborti M, Baig MJ. Characterization of rice straw from major cultivars for best alternative industrial uses to cutoff the menace of straw burning. Ind. Crop. Prod 2020; 143: 111919.
[26] Gullett B, Touati A. PCDD/F emissions from burning wheat and rice field residue. Atmospheric Environment 2003; 37: 4893-4899.
[27] Oanh NTK, Ly BT, Tipayarom D, Manandhar BR, Prapat P, Simpson CD, Liu LJS. Characterization of Particulate matter emission from open burning of rice straw. Atmospheric Environment 2011; 45: 493-502.

[28] Zhang Q, Zheng Y, Tong D, Shao M, Wang S, Zhang Y, Xu X, Wang J, He H, Liu W, Ding Y. Drivers of improved PM2.5 air quality in China from 2013 to 2017. Proceedings of the National Academy of Sciences 2019; 116 (49): 24463–24469.
[29] Al khalaf MN, Yousif HA, Use of rice husk ash in concrete. International Journal of Cement Composites and Lightweight Concrete 1984; 6(4): 241-248.
[30] Mistry R, Roy TK. Performance evaluation of bituminous mix and mastic containing rice husk ash and fly ash as filler. Con. Build. Mat. 2021; 268:121187. https://doi.org/10.1016/j.conbuildmat.2020.121187.
[31] Melotti R, Santagata E, Bassani M, Salvo M, Rizzo S. A preliminary investigation into the physical properties of biomass used as aggregate fillers for bituminous mixtures, Waste Management 2013; 33:1906-17. https://dx.doi.org/10.1016/j.wasman.2013.05.015.
[32] Sargın S, Saltan M, Morova N, Serin S, Terzi S. Evaluation of rice husk ash as filler in hot mix asphalt concrete, Con. Build. Mat. 2013; 48:390-97. http://dx.doi.org/10.1016/j.conbuildmat.2013.06.029.
[33] Al-Hdabi A. Laboratory investigation on the properties of asphalt concrete mixture with rice husk ash as filler, Con. Build. Mat. 2016; 126:544-51. http://dx.doi.org/10.1016/j.conbuildmat.2016.09.070.
[34] Tahami, SA, Arabani M, Mirhosseini AF. Usage of two biomass ashes as filler in hot mix asphalt, Con. Build. Mat. 2018; 170:547-556. https:// doi.org/10.1016/j.conbuildmat.2018.03.102.
[35] IS (Indian Standard). IS 2720: "Methods of test for soils First Revision". 1980. Bureau of Indian Standards, New Delhi, India.
[36] ASTM C 837-09: "Standard Test Method for Methylene Blue Index of Clay", West Conshohocken, PA, 2015.
[37] JTG E42-2005: "Standard Test Methods of Aggregate for Highway Engineering", Ministry of Transport 2005, Beijing, China.
[38] IS (Indian Standard).IS 1201-1220: "Method for Testing Tar and Bituminous Materials First Revision". 1978. Bureau of Indian standards, New Delhi, India.
[39] IS (Indian Standard). IS 73: "Paving Bitumen-Specification Fourth Revision". 2013. Bureau of Indian standards, New Delhi, India.
[40] IS (Indian Standard). IS 2386 (Part 1): "Methods of Tests for Aggregates for Concrete-Particle size and shape". 1963. Bureau of Indian standards, New Delhi, India.
[41] IS (Indian Standard). IS 2386 (Part 3): "Methods of Tests for Aggregates for Concrete – Specific Gravity, Density, Voids, Absorption and Bulking". 1963. Bureau of Indian standards, New Delhi, India.
[42] IS (Indian Standard). IS 2386 (Part 4): "Methods of Tests for Aggregates for Concrete – Mechanical Properties". 1963. Bureau of Indian standards, New Delhi, India.
[43] MoRT&H (Ministry of Road Transport and Highways). "Specifications for Roads and Bridges Fifth Revision". 2013. Indian Roads Congress (IRC), New Delhi, India.
[44] MS-2 (Manual Series): "Asphalt Mix Design Methods Seventh Edition". 2014. The Asphalt Institute, Lexington. KY, USA.
[45] ASTM D2493/2493M-16: "Standard Practice for Viscosity–Temperature Chart for Asphalt Binders". West Conshohocken, PA, 2016.
[46] ASTM D1075-07: "Standard Test Method for Effect of Water on Compressive Strength of Compacted Bituminous Mixtures". West Conshohocken, PA, 2007.
[47] AASHTO T283: "Standard Method of Test for Resistance of Compacted Asphalt Mixtures to Moisture-Induced Damage". 2002. American Association of Highway and Transportation Officials, Washington DC, USA.
[48] Gezencvej LB. Asphalt concrete for road surfaces. 343, Moscow:1985.
[49] Arabani M, Tahami SA. Assessment of mechanical properties of rice husk ash modified asphalt mixture. Con. Build. Mat. 2017; 149:350-358. https://dx.doi.org/10.1016/j.conbuildmat.2017.05.127.
[50] Gopalam J, Giri JP, Panda M. Effect of filler on bituminous base layer containing recycled concrete aggregates. International Journal of Transportation Science and Technology. 2020; 9: 239-248. https://doi.org/10.1016/j.ijtst.2020.04.002.

AUTHOR INDEX

Ajarostaghi, Seyed Soheil Mousavi 39, 55
Anirudh, N 127
Avirneni, Deepti 16
Bhojaraju, Chandrasekhar 16, 39, 55
Dhemla, Pankaj 77
Jayalekshmi, B R 71, 122
Kodeboyina, Ganesh Babu 16, 25
Krishna, A Shankara 1
Kumar, Aditya 115
Kuttagola, Imran 103, 115
Mane, Vishwanath 1
Manu 1
Mathews, Merin 71
Mousavi, Seyed Sina 39, 55
Nagarajan, Praveen 89
Nanjegowda, Vinay Hosahally 127
Prashanth, M H 103, 115
Raj, P K Arjun 89
Rao, Subba 1
Rathankumar, M N 127
Sarath, D 89
Satankar, Raj Kumar 77
Shashikala, A. P. 89
Somani, Prakash 77
Sreya, M V 122
Srinivasamurthy, Sharath 33
Swami, B L 77
Thejarathnam, T 103
Thotakura, Lakshmi 16
Venkataramana, Katta 71, 122